AF412522

The Release of
Neural Transmitter Substances

The Sherrington Lectures

In 1948, on the occasion of the ninetieth birthday of Sir Charles Scott Sherrington, O.M., G.B.E., M.D., F.R.S., the Council of the University resolved to institute a Lectureship in recognition of his distinguished contributions to Physiology and Medicine and of his association with the University of Liverpool as George Holt Professor of Physiology from 1895 to 1913.

The appointment to the Sherrington Lectureship is made biennially by the Council of the University on the joint recommendation of the Faculties of Medicine and Science.

The following lectures have been delivered and are published by Liverpool University Press:

I *Sensory Integration*, by E. D. Adrian, O.M., M.D., F.R.C.P., F.R.S., Professor of Physiology in the University of Cambridge.

II *The Frontal Lobes and Human Behaviour*, by John H. Fulton, O.B.E., M.D., D.SC., LL.D., Sterling Professor of the History of Medicine, Yale University.

III *The Invasive Adenomas of the Anterior Pituitary*, by Sir Geoffrey Jefferson, M.S.(LOND.), F.R.C.S., F.R.S., Emeritus Professor of Neuro-Surgery, University of Manchester.

IV *Sherrington: Physiologist, Philosopher and Poet*, by the Right Honourable Lord Cohen of Birkenhead, M.D., D.SC., LL.D., F.R.C.P., F.A.C.P., F.F.R., F.S.A., J.P., Professor of Medicine in the University of Liverpool.

V *The Excitable Cortex in Conscious Man*, by Wilder Penfield, O.M., C.M.G., LITT.B., M.D., D.SC., F.R.C.S., HON.F.R.C.P., F.R.S., Director, Montreal Neurological Institute, McGill University, Montreal, Canada.

VI *Visual Pigments in Man*, by W. A. H. Rushton, SC.D., F.R.S., Reader in Physiology in the University of Cambridge.

VII *The Conduction of the Nervous Impulse*, by A. L. Hodgkin, SC.D., F.R.S., Foulerton Research Professor of the Royal Society, University of Cambridge.

VIII *The Cerebral Control of Movement*, by Derek Denny-Brown, O.B.E., M.D., D.PHIL., LL.D. (HON.), James Jackson Putnam Professor of Neurology, Harvard University.

IX *The Inhibitory Pathways of the Central Nervous System*, by John C. Eccles, F.R.S., M.B., B.S., N.A., D.PHIL., F.R.A.C.P., F.R.F.N.Z., F.A.A., Faculty of Health Services, State University of New York at Buffalo.

X *The Release of Neural Transmitter Substances*, by Bernard Katz, M.D., D.SC., F.R.S., Professor of Biophysics, University College London.

The Sherrington Lectures X

The Release of Neural Transmitter Substances

Bernard Katz, M.D., D.Sc., F.R.S.

Professor of Biophysics,
University College London

CHARLES C THOMAS · PUBLISHER
Springfield · Illinois · U.S.A.

CHARLES C THOMAS · PUBLISHER
BANNERSTONE HOUSE
301-327 East Lawrence Avenue, Springfield, Illinois, U.S.A.

Published simultaneously in the British Commonwealth by
LIVERPOOL UNIVERSITY PRESS, LIVERPOOL, ENGLAND

Published simultaneously in Canada by
THE RYERSON PRESS, TORONTO

*No part of this book may be reproduced in any form
without permission from the publisher, except for
the quotation of brief passages in criticism*

Copyright © 1969 by Bernard Katz

Printed in England by Hazell Watson & Viney Ltd., Aylesbury, Bucks

Preface

This monograph is based on a Sherrington Lecture which I had the honour to give at Liverpool in October 1967. I should like to record here my deep gratitude to the colleagues who invited me, and in particular to Professor R. A. Gregory, F.R.S., for much help and encouragement.

In preparing the manuscript I have adhered closely to the line of argument which I followed during the lecture itself, expanding a little here and there, but avoiding side-issues, at least in the main body of the text. I have made up for this by adding a separate 'commentary', a kind of appendix in which I have discussed several items, some of them of a more controversial nature, which were outside the direct line of the lecture.

Finally, I should like to emphasize what will be obvious to those familiar with the field, namely that this lecture is a progress report on many years of collaborative work, in which my colleagues, Paul Fatt, José del Castillo, Ricardo Miledi, and I, have joint responsibility.

B.K.

Contents

Introduction   1

The neuromuscular junction   2

Spontaneous electric activity at the motor end-plate   5

Quantal release of acetylcholine from motor nerve endings   9

Acetylcholine release—an electrically controlled form of secretion   11
The end-plate potential and its quantal components   12
A working hypothesis   14
Statistical properties of transmitter release   16

Recording from small parts of a myoneural junction   19

The use of the focal calcium pipette   22
Synaptic delay and time course of transmitter release   25
More about Poisson distributions   28

Junctional transmission without impulses   30

The role of calcium in junctional transmission   33
The giant synapse of the squid   35

Commentary   41

A comment on terminology   41
Variability of the quantal packet   42
Quantal and non-quantal release   43
Is the release of acetylcholine an 'auto-catalytic' process?   43
Repetitive 'back-firing' in mammalian nerve-muscle preparations   45
Generality of quantal transmitter release   46
Structural and biochemical evidence related to quantal release of acetylcholine   47

Bibliography   49

Author Index   57

Subject Index   59

Figures

1 Photomicrograph of neuromuscular junction   3

2 Electron micrograph of neuromuscular junction   4

3 Spontaneous miniature end-plate potentials (min. e.p.p.'s)   6

4 Effect of curare   7

5 Effect of edrophonium   8

6 Electrical control of frequency of min. e.p.p.'s   12

7 Circuit equivalent of min e.p.p.'s   13

8 Step-wise fluctuations of end-plate responses   14

9 Vesicular hypothesis of quantal release   16

10 Statistical analysis of end-plate response   18

11 Application of Poisson's theorem to end-plate response   19

12 Focal recording of min. e.p.p.'s   21

13 The use of the calcium pipette   23

14 Impulse propagation in motor nerve endings   24

15 Focal recording of pre- and post-synaptic response at the neuromuscular junction   25

16 Distribution of synaptic delays   26

17 Effect of temperature on synaptic delay   27

18 Temporal dispersion of quantal transmitter release   28

19 Transmitter release in muscle paralysed by tetrodotoxin   32

20 Effect of calcium on transmitter release   34

21 The 'latency shift' at the neuromuscular junction   36

22 Input/output relation from the giant synapse of the squid   37

23 Suppression of transmitter release during presynaptic potential change   38

The Release of
Neural Transmitter Substances

INTRODUCTION

The purpose of this lecture is to discuss the mechanism by which a nerve axon releases a transmitter substance from its terminals. Most of the evidence which I shall present has been derived from physiological experiments on the neuromuscular junction, and the reader might therefore feel that this is not sufficient to justify the generalization implied in my title. In answer to this objection, I think it has become very clear in recent years that the basic mechanisms revealed by a study of neuromuscular transmission are the same as those underlying synaptic transmission, at least at those synapses in the central and peripheral nervous system at which chemical mediation operates.

When Sherrington developed the concept of a synapse, that is a functional connexion between two separate nerve cells, he did not include the neuromuscular junction. But he did comment on the structural and functional similarities between interneuronal contacts and those junctions which nerve endings make in the periphery, with muscles and other effector organs. Sherrington pointed out that in all these cases we have to do with the transfer of signals between separate cells, separated by their membranes, and that the processes involved in intercellular conduction are likely to differ in many important respects from the process of impulse conduction within the confines of a single neuron.

This difference was brought out most strikingly when H. H. Dale (1936) and his colleagues, G. L. Brown (1936), W. Feldberg and Marthe Vogt, showed that the action potential wave in the motor nerve is not transferred directly to the muscle fibre but produces its excitatory effect by the liberation of a chemical stimulant from the nerve ending. Between the nerve impulse (N) and the initiation of an action potential wave in the muscle fibre (M), a chemical mediator is interposed—which in this particular instance has been identified as acetylcholine (ACh).

$$N \rightarrow \underset{\text{I}}{ACh} \rightarrow \underset{\text{II}}{M}$$

The problems which my colleagues, Paul Fatt, José del Castillo, Ricardo Miledi and I have been studying over the last 17 years, are those concerning the two intermediate mechanisms, I and II. The process by which acetylcholine gives rise to a muscle impulse can now be described in terms of a membrane reaction which leads to a non-selective increase of cation permeability (mainly to sodium and potassium, but also to calcium and a variety of quaternary ammonium ions). This in turn produces a local depolarization of the muscle fibre, known as the 'end-plate potential'. Normally the reaction is so intense that the resulting end-plate potential rises at a high rate and rapidly exceeds the excitation threshold at which an impulse is initiated. The new impulse travels along the whole length of the muscle fibre and activates the contractile mechanism at every point.

It may be noted, in passing, that the study of the end-plate potential and of its chemical origin gives us important clues to the much wider problem of biological 'chemo-electric transducers' which are the basic elements on which the function of chemical sense organs, and chemo-receptor systems generally, depends. The ionic mechanism of the end-plate potential has been dealt with in several reviews (del Castillo and Katz, 1956a; Nastuk, 1959; Katz, 1958, 1962; Ginsborg, 1967), and I do not propose to discuss it in the present lecture. I want to concentrate on step I, that is the 'secretory' function of the motor nerve ending, and the mechanism by which a nerve impulse manages to accelerate the rate of acetylcholine secretion.

THE NEUROMUSCULAR JUNCTION

A few words should be said about the experimental preparation and the physiological techniques. Much of this work was done by intracellular electrical recording from the junctional region of frog muscle fibres. We usually call this region a motor 'end-plate', although strictly speaking this term applies only to the plate-shaped junctions in certain classes of animals (lizards, mammals). Fig. 1 shows a photomicrograph of a myoneural junction which was fixed and stained by R. Miledi after an experiment. Two kinds of stain were used: Sudan black to show the medullated axon which approaches the muscle fibre and, in this particular case, divides into two terminal myelin segments, and a special-histo-chemical technique for locating cholinesterase (Koelle and Friedenwald, 1949; Couteaux, 1955, 1963) by which the products of hydrolysis of acetylthiocholine are precipitated and shown up as a dark brown stain. This is nowadays the most reliable histological method of demonstrating motor end-plates, and it rests on the fact that a potent enzyme which hydrolyses acetylcholine is concentrated in the immediate

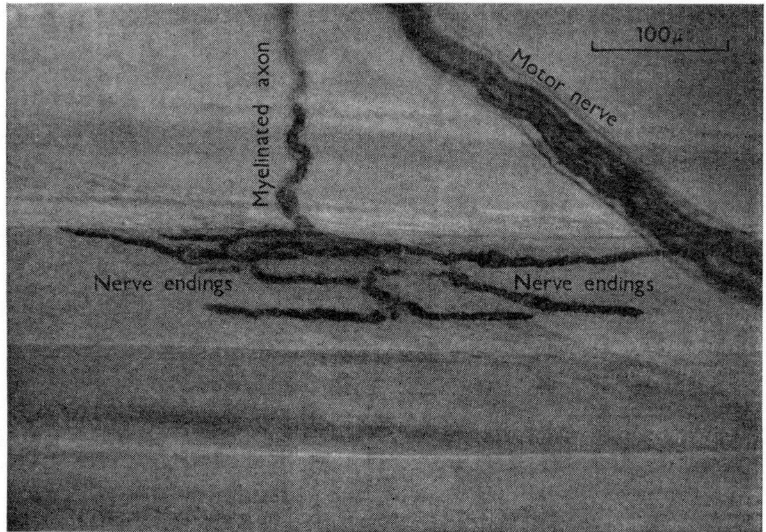

Fig. 1. Photomicrograph of a neuromuscular junction in frog sartorius. 'Cholinesterase stain' was used to show up the non-myelinated branchwork of the nerve endings. In addition, Sudan black was used for the myelinated parts of the nerve.

surroundings of the non-myelinated motor nerve terminals. We have known, of course, since the work of Dale and his colleagues that the principal function of this enzyme is to curtail the local action of acetylcholine and, after each nerve impulse, to ensure a rapid return of the system to its resting state. It should perhaps be pointed out that the position of the non-myelinated terminals is not readily visible in the living, unstained preparation. All one can see there are the ultimate segments of the myelinated axon on the surface of a muscle fibre. But as I shall explain later, we can obtain electrophysiological clues which help us to map out the location of the synaptic terminals on the living muscle fibre, and we can check our result by staining the preparation after the experiment.

In Fig. 2, a small part of the neuromuscular junction is shown at higher magnification with the help of the electron microscope. We see the muscle fibre with its sarcomeres and longitudinal filaments, its surface being thrown into a regular array of transversely aligned 'postsynaptic folds'. The muscle fibre is separated from the axon terminal by an extracellular synaptic cleft. (This space becomes filled with precipitate during the 'cholinesterase staining' method, see Couteaux, 1963;

Lewis and Shute, 1964; Miledi, 1964). Note that the so-called 'basement membrane' follows the contours of the muscle fibre within the synaptic cleft and even into the junctional folds. The nerve ending is overlaid by a Schwann cell layer which separates it from the superficial connective tissue. The subcellular structure of the 'presynaptic' terminal is of great interest: there are numerous mitochondria near the Schwann cell layer,

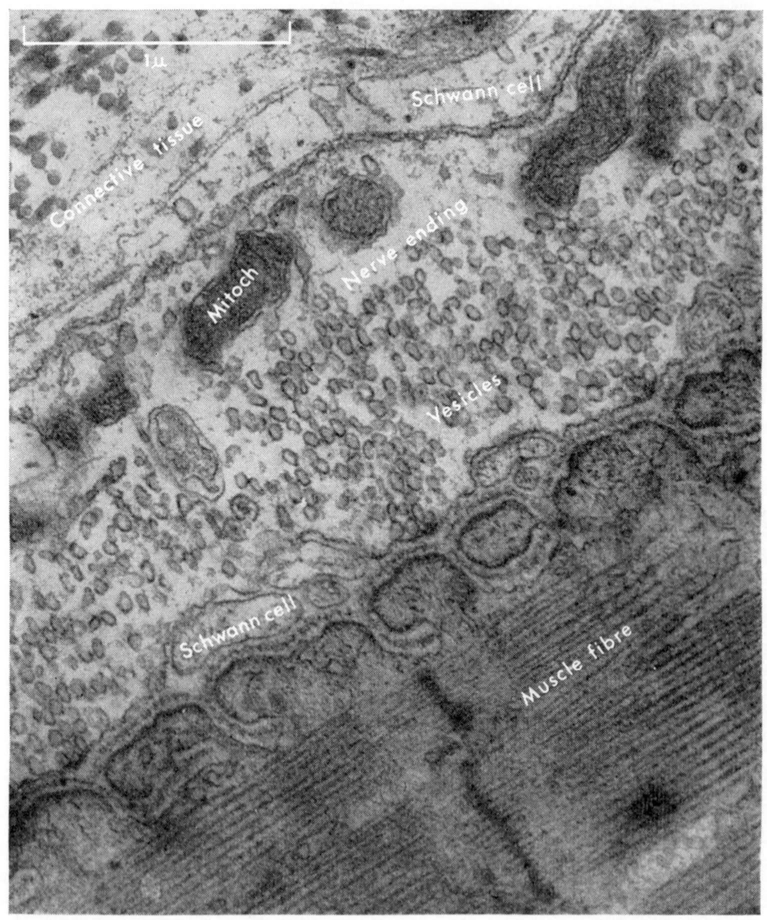

Fig. 2. Electron micrograph of part of myoneural junction from frog sartorius. Longitudinal section of muscle. Note: presence of mitochondria and synaptic vesicles inside the nerve ending, and 'basement membrane' in the synaptic cleft and postjunctional folds. (From Birks, Huxley and Katz, 1960.)

4

and a characteristic accumulation of 'synaptic vesicles' is seen on the synaptic side. These vesicles tend to form clusters which aggregate at certain points of the axon membrane, usually at a region which faces a postsynaptic fold.

The experimental methods are developments of the micro-electrode technique, introduced by Graham and Gerard (1946) and Ling and Gerard (1949). Very fine glass pipettes drawn out to form tips of less than 0.5μ outside diameter, and filled with a concentrated KCl solution, can be used to penetrate the muscle membrane without causing appreciable damage, and to record the potential difference across the membrane directly, against a large reference electrode in the outside bath. It is possible to use the same kind of pipette on the surface of the junction to record focal potential changes which occur at localized regions of the nerve or muscle membrane. Finally, micro-pipettes of this kind can be filled with concentrated drug solutions and, by close-range ionophoretic application, can be used to test the pharmacological properties of the muscle fibre and its end-plate region.

SPONTANEOUS ELECTRIC ACTIVITY AT THE MOTOR END-PLATE

If one penetrates the muscle fibre with a micro-electrode, one finds that the inside is at a negative potential, about 90 mV with respect to the outside bath. The same result is obtained wherever one inserts the micro-pipette, in the vicinity of the nerve contacts as well as at all other points of the fibre. But if one watches the oscilloscope at high amplification, one finds that, unlike the rest of the muscle fibre, the junctional region is in a state of subthreshold electric activity even when no stimulus has been applied and when—to all outward appearance—the muscle fibre is completely at rest (Fig. 3). One observes a sequence of small, intermittent electric discharges, each of them amounting to a brief depolarization of about 0.5 mV, rising rapidly and declining more slowly, with a total duration of about 20 msec. The potential changes are very similar to the end-plate potential which the nerve impulse produces in the muscle fibre, the main difference being their much smaller amplitude (about 0.5 instead of over 50 mV) and their 'spontaneous' occurrence. Paul Fatt and I called them 'miniature end-plate potentials' (1952).

If one examines the sequence of time intervals between successive discharges, one usually finds them distributed in a completely random fashion. Occasionally there are transient bursts of high-frequency discharges, but more often there is no sign of mutual interaction, nor is there any refractory period or 'dead time'.

In general, in the absence of an applied stimulus, miniature end-plate potentials (min. e.p.p.'s) occur at random intervals, with a *probability* which remains constant over long periods of time and is independent of previous occurrences.

P. Fatt and I suspected that the min. e.p.p.'s are caused by leakage of acetylcholine from the nerve ending. The motor axon contains an enzymic apparatus which continuously builds up and replenishes a store

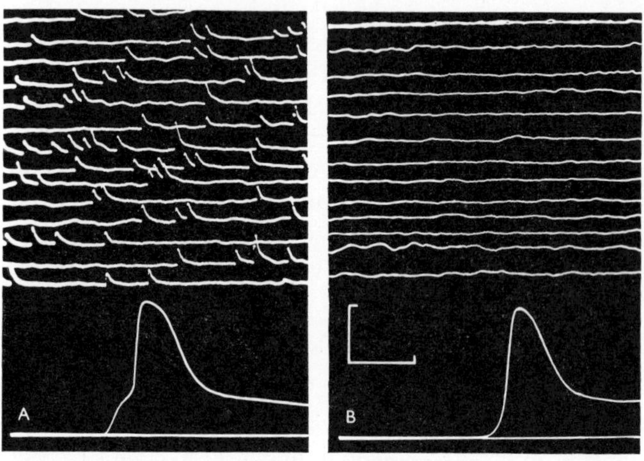

Fig. 3. Spontaneous 'miniature end-plate potentials'. (From Fatt and Katz, 1952.) A: intracellular recording at an end-plate. B: recorded 2 mm away in same muscle fibre. Upper portions were recorded at low speed and high amplification (calibrations 3·6 mV and 46 msec): they show the localized spontaneous activity at the junctional region. Lower records show the electric response to a nerve impulse, taken at high speed and lower gain (calibrations 50 mV and 2 msec). The stimulus was applied to the nerve at the beginning of the trace; response A shows step-like end-plate potential leading to a propagating muscle spike; in B, the spike alone is recorded after additional delay due to conduction along 2 mm of muscle fibre. In all figures, unless otherwise stated, upward deflexion means positive-going potential change at the micro-electrode.

of acetylcholine in the synaptic terminals, and one might expect the transmitter substance to 'overspill' and to be released at a low rate even during the absence of any impulse traffic. The postsynaptic muscle membrane is highly sensitive to acetylcholine, so that the local impact of even quite small amounts of this substance would be registered in the form of a transient potential change.

This interpretation was greatly strengthened by the following experiments. It is possible to imitate the miniature potential by closely approaching the end-plate region with a micro-pipette which contains a solution of acetylcholine (ACh) chloride, and by discharging a small quantity of ACh ionophoretically. This produces a local depolarization of the muscle fibre (see Figs. 4 and 5). The response to the ACh can be modified by a number of specific drugs: thus, curare suppresses it, while inhibitors of cholinesterase (e.g. neostigmine, edrophonium) cause it to increase in amplitude and duration. By using an assembly of drug-filled

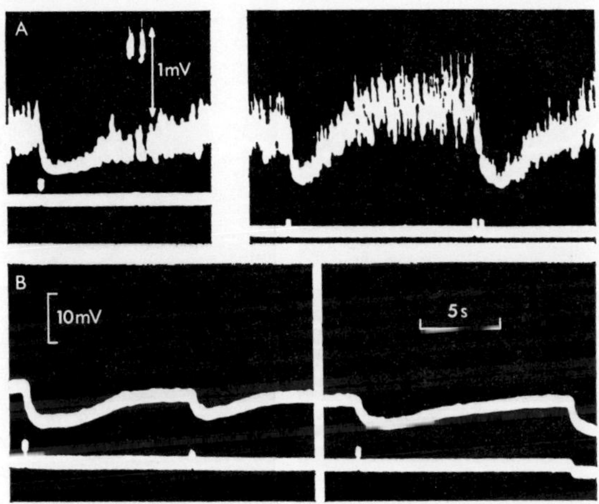

Fig. 4. The effect of ionophoretic doses of curarine (A): on massive bursts of min. e.p.p.'s; (B) on a steady depolarization produced by locally applied ACh. At end of record (B), ionophoretic ACh application was stopped. In each block of records, the upper trace shows membrane potential change, the lower trace monitors ionophoretic currents (calibration: '10 mV scale' $= 1 \cdot 1 \times 10^{-7}$A). (From del Castillo and Katz, 1957b.)

micro-pipettes one can subject the spontaneously occurring miniature end-plate potentials to stringent tests and see whether their pharmacological reactions to curare and anti-cholinesterases match those of the artificially produced ACh-potential.

Such tests are illustrated in Figs. 4 and 5. In Fig. 4 the curare-pipette had been used to 'prod' a superficial nerve ending, and after some local mechanical 'irritation', a prolonged high-frequency outburst of miniature potentials occurred. The individual discharges were so frequent that they summated and caused the baseline to be raised by nearly 1 mV above the resting level. A brief pulse of tubocurarine promptly reduced the size of the potentials. In the lower part (Fig. 4B), ACh was allowed to diffuse from the tip of another micro-pipette and caused a steady depolarization of several millivolts (terminated at the end of the record). It will be seen that the pulses of curare have very similar suppressing effects on the summating min. e.p.p.'s (upper), and on the artificially produced ACh-potentials (lower records).

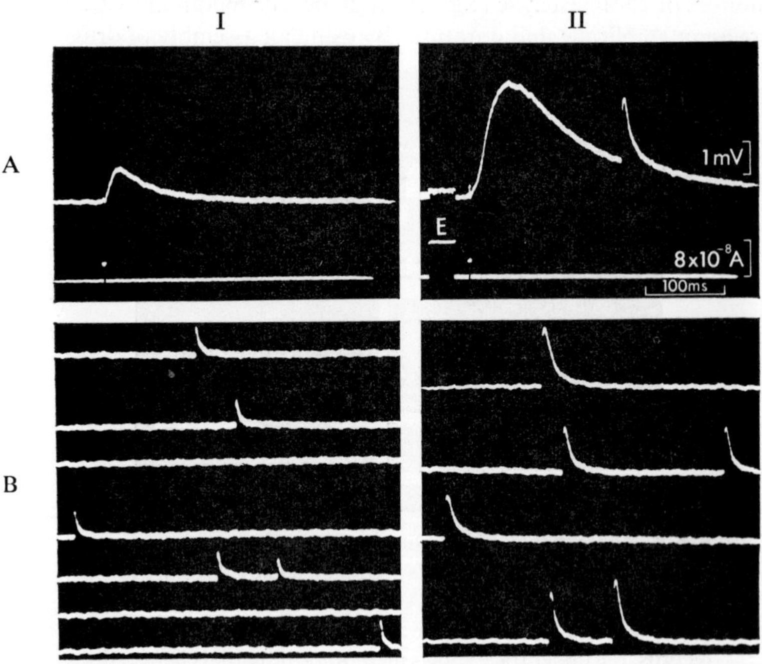

Fig. 5. The potentiating effect of edrophonium, (A): on ACh-potential; (B): on min. e.p.p.'s. (From an experiment by R. Miledi.) In (A), a brief pulse of ACh was applied, alone (I), and after a pulse of edrophonium (E, II). Pulses are monitored in lower traces. In IIA, a large spontaneous potential is seen during the decline of the ACh-potential. In (B): min. e.p.p.'s are recorded. (I) before, and (II) during, steady ionophoretic application of edrophonium.

This experiment suggests that the miniature potentials are due to the spontaneous impact on the junctional region of the muscle fibre of some substance which acts like ACh and whose effect is antagonised, like that of ACh, by curare. This test, however, does not distinguish between ACh and a variety of other depolarizing agents, e.g. various stable choline esters to which the end-plate region is known to be highly sensitive. The experiment shown in Fig. 5 takes the analysis a step further by comparing the effects of a fast-acting cholinesterase inhibitor (edrophonium) on ACh- and miniature end-plate potentials.

Edrophonium, like neostigmine and other anti-esterases, potentiates the ACh-effect (Katz and Thesleff, 1957b). This must be ascribed to a protection of ACh against enzymic hydrolysis, because no such potentiation is observed if, instead of ACh, stable choline esters like carbachol or decamethonium are used. It is evident from Fig. 5 that the spontaneous miniature potentials (lower records) are increased in size and duration by edrophonium, in a way quite similar to the ACh-potential (upper records). In its pharmacological reactions to specific suppressing and potentiating drugs, our spontaneous miniature potential resembles very closely the much larger end-plate potential which the nerve impulse produces in the muscle fibre, and there is little doubt that both are due to local impacts of acetylcholine upon the chemo-sensitive part of the postsynaptic membrane.

QUANTAL RELEASE OF ACETYLCHOLINE FROM MOTOR NERVE ENDINGS

The spontaneous potentials share many properties with the artificial ACh-potential, but there is an important difference. Miniature potentials are discrete all-or-none events, of standard size and time course. The ACh-potential, however, can be continually reduced in size below a detectable level and it can be changed in its time course smoothly, by adjusting the strength and duration of the ionophoretic pulse and by simply varying the distance between the pipette and the end-plate. This shows that the individual molecular components of the acetylcholine action cannot be resolved by our method. Although the end-plate is a very sensitive electric indicator for small doses of ACh, it is not sensitive enough to show up the action of single molecules. The continuous smooth gradation of the ACh-potential indicates that the electrical effects of molecular collisions are far below the noise level of our recording system. And it follows that a discrete potential change like a min. e.p.p. cannot be explained by random escape of one or a few molecules of ACh from the nerve ending, but must be due to a synchronous impact of a large multi-molecular packet of ACh.

We should like to know the precise molecular content of this quantal packet, but this remains quite uncertain and we cannot even state the order of magnitude with confidence. Attempts to assay it by close ionophoretic application are bound to give us an overestimate. R. Miledi

(1961) found that under optimal conditions, approximately 10^5 molecules of ACh must be discharged from the pipette to produce a depolarization whose potential x time area is equivalent to a miniature potential. This would represent an upper limit: even the best micro-pipetting technique is not likely to be as effective as the nerve ending in applying a dose of ACh to the receptors in the synaptic grooves and folds, and we may surmise that, unlike the 'natural product', most of the ionophoretically released ACh diffuses away to the outside without reacting with the muscle membrane.

A lower limit of the packet size could perhaps be arrived at by the following argument: In *small* muscle fibres, one can record miniature potentials of several millivolts amplitude. (The smaller the muscle fibre, the larger is the size of the miniature potential. This arises from the fact that the potential change in different fibres originates from an approximately constant amount of ionic flux; the voltage change therefore becomes greater the higher the characteristic input resistance of the fibre (Katz and Thesleff, 1957a)). Under optimal recording conditions, one can observe miniature potentials which are nearly 100 times larger than the noise fluctuations of the amplifier input. If a miniature potential of a signal/noise ratio of 100 were produced by only 100 molecules, then a minimal dose of ACh applied to a small fibre would be expected to show molecular effects in the form of increased noise fluctuations. This is a point which deserves further study.

As regards the molecular size of the quantal packet, we are for the present left with a range of uncertainty starting, I think, from more than 100 and going up to less than 100,000 molecules! I suspect the right order of magnitude is somewhere in the middle.

Can these estimates be reconciled with the idea (see e.g., del Castillo and Katz, 1955b) that the quantal packets have their structural basis inside the synaptic vesicle? The suggestion is plausible, for the vesicles are found inside the nerve ending aggregated at the very sites where transmitter release is presumed to occur (Fig. 2). To fit 10^3 to 10^4 molecules of ACh into each of these small 500 Å bags would not be too difficult, though 10^4 molecules would give us a hypertonic solution; 10^5 molecules would be stretching our imagination a bit too far.

I have presented evidence for the view that the spontaneous potentials arise from a random impact of large multi-molecular packets of ACh upon the receptive region of the muscle membrane. There is little doubt that the ACh comes from the outside, and not from within the muscle fibre, for if one discharges ACh ionophoretically into the interior of the fibre, no effect is observed (del Castillo and Katz, 1955a). The whole

10

packet of ACh reaches the end-plate almost synchronously (the ionic current rises to a peak within a fraction of a millisecond, see Fatt and Katz, 1952; Katz and Miledi, 1965b) which tells us that the release occurs nearby, in the immediate neighbourhood of the end-plate. All this, of course, fits in with the idea that the ACh-packets are discharged spontaneously by the nerve terminal which is known to be a rich source of the substance (de Robertis *et al.*, 1961; Whittaker *et al.*, 1964).

There are several pieces of evidence which strongly support this view. If one sections the motor nerve, the terminals degenerate after some days and at that time the spontaneous discharge stops while there is no change in the local sensitivity of the muscle fibre to ACh (Birks, Katz and Miledi, 1960; Miledi and Slater, 1963). Later on, spontaneous discharges may reappear at very low frequency. This occurs only in certain species and is probably due to the activity of Schwann cells which can replace the nerve axon in its terminal synaptic position. Another significant finding was that botulinum toxin which specifically interferes with the release of transmitter from cholinergic nerves (Burgen, Dickens and Zatman, 1949), also causes a stoppage of min. e.p.p.'s (Brooks, 1956; Thesleff, 1960). Finally, it has been shown that the frequency of the discharge can be 'driven' over a wide range by altering the membrane potential in the terminal part of the motor axon (but not by changing the membrane potential of the muscle fibre).

Acetylcholine release—an electrically controlled form of secretion

If one places two electrodes along the pre-terminal part of the motor axon with the cathode near the junction, a steady current through the electrodes causes the terminal part of the axon to become depolarized. This is associated with a rise in the frequency of the miniature potentials (del Castillo and Katz, 1954c; Liley, 1956c; see Fig. 6). By increasing the current strength, the rate of the discharge can be raised more than a hundredfold until the potentials become so frequent that they can no longer be counted individually.

This is a very significant finding. We have here an electrically controllable rate of ACh secretion, and there is now conclusive evidence that the same process is put into operation—at much higher intensity and for a much shorter time—when the action potential wave arrives in the motor endings.

The nerve impulse does not initiate a new process, but accelerates the rate of secretion of ACh packets by a factor of a few hundred thousand for a very brief interval of time. As a result, instead of an intermittent discharge at a mean rate of about one per second, a few hundred packets are released almost synchronously, within less than one millisecond, and produce the large end-plate potential which immediately starts up an impulse in the muscle fibre.

It should be stated, in passing, that min. e.p.p.'s do not go on adding their amplitudes in proportion to their numbers, except over a very

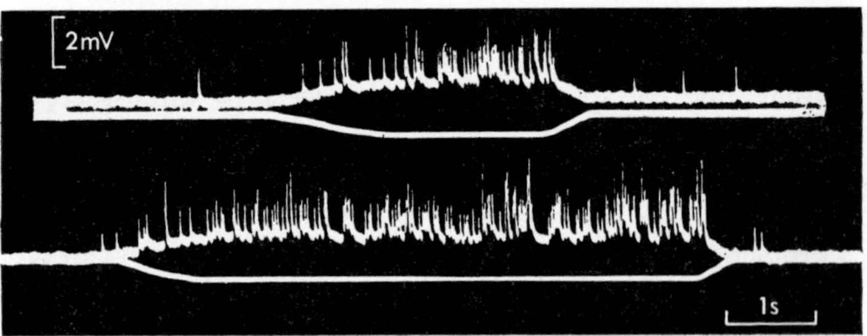

Fig. 6. Electrical control of the frequency of min. e.p.p.'s (cf. del Castillo and Katz, 1954c). In each pair of traces, the upper shows miniature potentials, the lower indicates current flowing through the terminal part of the motor axon. The cathode was placed near the junction so as to depolarize the nerve endings.

small range. This is because the potential changes arise from parallel increments in membrane conductance (see Fig. 7), and even if the conductance were to grow to infinity, it could lead only to a depolarization of some 70 to 80 mV (del Castillo and Katz, 1954 d). What *is* additive in strictly linear fashion, is the change in ion conductance or permeability, produced by each packet of transmitter. So long as the number of packets is small, approximately linear summation of amplitudes will occur, but if the packets become numerous and the depolarization exceeds a few percent of the resting potential, each additional unit of conductance increase produces a progressively smaller increment of the e.p.p. (del Castillo and Katz, 1954b; Martin, 1955, 1966).

The end-plate potential and its quantal components

That the end-plate potential, i.e, the postsynaptic response to a motor impulse, is made up of quantal components each of which is identical with the spontaneous miniature potential, became clear in a series of experiments in which the calcium concentration of the Ringer solution

12

was altered. It has been known since the days of Sydney Ringer and F. S. Locke (1894) that a small quantity of calcium in the muscle bath is necessary for the maintenance of neuromuscular transmission. We know now that calcium is required for the process of transmitter release, by an impulse or by artificially applied depolarization (Harvey and MacIntosh, 1940; del Castillo and Katz, 1954b, c; Boyd and Martin, 1956b; Liley, 1956b, c; Katz and Miledi, 1965c) and that magnesium competes with calcium and acts as an inhibitor of the release mechanism (Jenkinson, 1957; Dodge and Rahamimoff, 1967). It should be added that while these ions have a very powerful influence on the secretory response of the nerve ending to depolarizing stimuli, they have relatively little effect

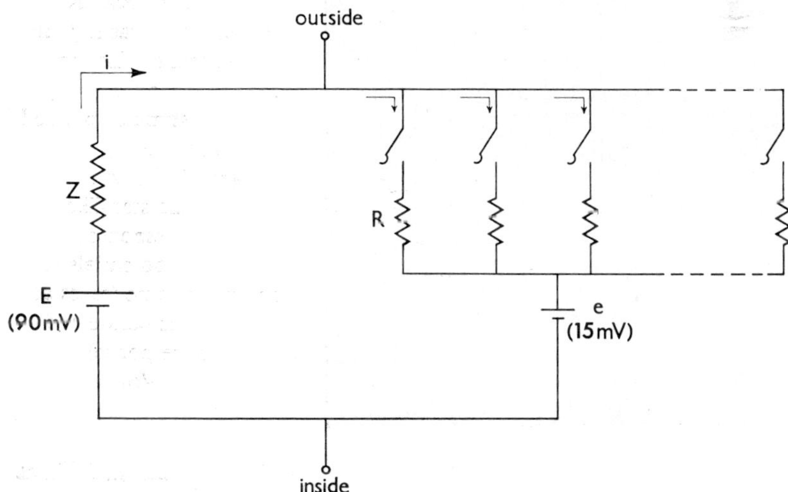

Fig. 7. Diagram illustrating electric circuit equivalent of min. e.p.p.'s. The resting potential of the muscle fibre is represented by E, and the characteristic 'input impedance' of the resting fibre by Z (of the order of 0·1 megohm). Each min. e.p.p. arises from the temporary closure of one key, which shunts the fibre membrane locally by a resistance R (of the order of 10 megohm) in series with a small voltage e (15 mV, negative inside). The result is flow of depolarizing current i, and a small reduction of the p.d. between inside and outside. Closure of additional keys produces larger depolarization, but the increments become progressively smaller the larger the number of keys (n), and the depolarization cannot exceed (E − e), i.e. 75 mV. But as long as R/n ≫ Z (i.e. if n is sufficiently small), the min. pots. add nearly linearly.

on the rate of the spontaneous discharge, in the absence of a depolarizing agent (Fatt and Katz, 1952; del Castillo and Katz, 1954a; Boyd

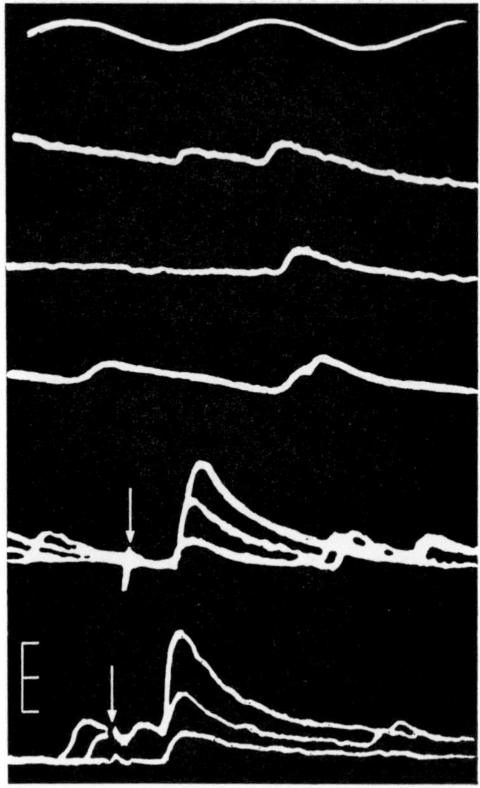

Fig. 8. Step-wise fluctuations of e.p.p. responses during calcium deficiency. (From Fatt and Katz, 1952). The upper three records show spontaneous potentials. The frequency was high in this preparation. Lower records show e.p.p. responses (stimulus marked by arrow, three super-imposed traces in each, showing step-like variation in response; spontaneous potentials are seen on the same traces at random times before and after e.p.p. response). Time: 50 c/s. Volt scale: millivolts.

and Martin, 1956a; Liley, 1956a; Hubbard, 1961; Miledi and Thies, 1967).

When calcium is withdrawn (or magnesium added), the end-plate potential is greatly reduced in size and eventually vanishes. If one watches the process by recording from a single end-plate, it becomes clear that this is not a smooth and gradual decline, but the response diminishes discontinuously and fluctuates in what appear to be quantal steps. Fig. 8 shows an example from an experiment by Dr. P. Fatt and myself: it looks as though a small fluctuating number of miniature potentials make up successive e.p.p.'s.

A working hypothesis

At this stage, I should like to introduce a working hypothesis which my colleagues and I have found useful in planning our experiments. We

14

assume that each unit packet of ACh (whose release produces a miniature potential) is pre-formed within a synaptic vesicle in the nerve terminal. The vesicle may be supposed to accumulate the transmitter substance actively and maintain it at a high concentration, much higher than exists in the surrounding axoplasm. Whether the transmitter is synthesized inside the vesicle (de Robertis *et al.*, 1961) or in the cytoplasm, and merely accumulated to a high concentration and stored for subsequent release by the vesicle, is a question which need not concern us at the moment.

The transmitter substance, parcelled up inside a vesicular bag, is separated from its postsynaptic target by two barriers, the vesicular membrane itself and the membrane of the axon terminal. The transmitter must be released in such a way that the whole packet reaches the receptor membrane almost synchronously and at high concentration. One can rule out as unlikely any mechanism that does not simultaneously lead to the opening of the two intervening membrane barriers. For example, discharge of the vesicular content into the axoplasm, or extrusion of intact vesicles into the synaptic cleft, are *a priori* unlikely propositions. Release could result from a collision between vesicular and axon membranes. Now I envisage that the frequency of collisions is very high at all times, but the vast majority of such collisions are unsuccessful and do not lead to transmitter release. Release would occur only in the statistically improbable event of two reactive molecules or specific sites, one on each membrane, meeting and starting off a reaction which causes the colliding barriers to burst open (Fig. 9). In less vulgar language this process is known as 'exocytosis'. Ultrastructural evidence has been obtained in several instances, e.g., for the secretion of zymogen granules by Palade (1959) and in the case of neurosecretory activity by Smith and Smith (1966). There are suggestions that this process is rapidly reversible so that a vesicle—having discharged its contents— would quickly detach itself from the membrane and start re-accumulating the transmitter substance.

Depolarization, whether by an applied current or during impulse invasion, produces important changes in the axon membrane, but does not significantly or immediately disturb the axoplasm and its enclosed organelles. Our suggestion is that the nerve impulse changes the axon membrane so that it presents temporarily many more reactive sites to the colliding vesicles, with the result that the statistical probability of release is increased, and that for the same total number of collisions a much larger fraction becomes successful. This, for the moment, is our working hypothesis; let us see how we get on with it.

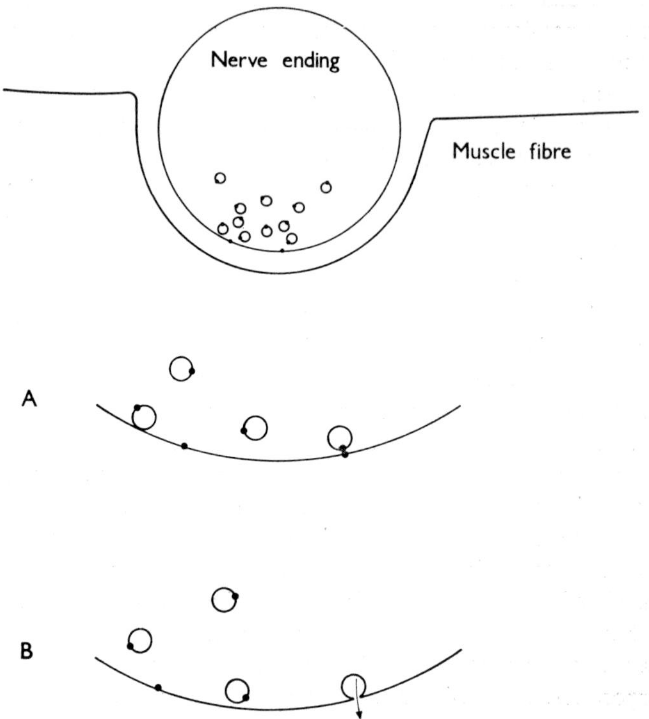

Fig. 9. (From del Castillo and Katz, 1957a). Schematic diagram to explain quantal release of transmitter, as a result of critical collision between vesicle and axon membrane, leading to temporary membrane fusion and 'all-or-none' discharge of vesicular contents into the synaptic cleft. Reactive sites on the surface of the vesicles and of the axon membrane are shown as black dots. Release of transmitter occurs only when, in the process of vesicular collision with the axon membrane, the two sites meet. The collision rate is assumed to be high, but 'successful' collisions are statistically rare. Their probability of occurrence varies directly with the number of reactive sites in the axon membrane.

Statistical properties of transmitter release

The situation then is that we have a large population of ACh packets milling about inside the terminal, queuing up for release by the next impulse, but the chance of success for any individual is very low indeed.

16

In a situation of this kind, Poisson's statistical law should apply; that is to say, if we know the average number m of packets released by an impulse, we can predict the *probability* of obtaining 0, 1, 2, 3 ... x packets, from the equation

$$p_x = e^{-m}m^x/x!$$

Thus, if we know m, Poisson's theorem tells us how often we are likely to encounter responses made up of 0, 1, 2, 3 etc., packets. If the average number m is small, there will be an appreciable proportion of failures, i.e., zero e.p.p.'s, given by

$$p_0 = \text{(Number of e.p.p. failures)}/\text{(Number of impulses)} = e^{-m}.$$

Hence, one way of determining m is simply to count responses and failures in a large series of observations, from which we obtain

$$m = ln[\text{(No. of impulses)}/\text{(No. of failures)}] \qquad (1)$$

Now, we have an independent way of measuring m, and this provides a critical test for the hypothesis, viz., that the release of transmitter by the nerve impulse is 'quantal', and that it involves the same unit packets as those which are secreted spontaneously, Provided the e.p.p. is small, and we are working within the range of linear summation of min. e.p.p.'s, the value of m should be given by the following ratio:

$$m = \text{(mean size of e.p.p.)}/\text{(mean size of spont. min. e.p.p.)} \qquad (2)$$

This equation merely restates our hypothesis, and all we have to do is measure average amplitudes (including zero responses) in a large series of observations, which is a good procedure with a high degree of built-in accuracy.

The consistency of these two equations, (1) derived from response and failure counts, and (2) from measurements of mean amplitudes, has been tested by several groups of investigators, on frog and mammalian nerve-muscle junctions (del Castillo and Katz, 1954b; Boyd and Martin, 1956b; Liley, 1956b) and excellent agreement has been found over the whole range of m values (0 to 3) in which a sufficient number of failure counts occur.

Poisson's theorem also tells us how many of our e.p.p.'s are likely to be made up of 1, 2, 3, etc., miniature components. Having counted the number of failures (n_0), it is an easy matter to calculate the expected number of single, twin, triplet responses etc. in the following way:

$$n_0 = e^{-m}; \qquad n_1 = mn_0; \qquad n_2 = \frac{m}{2}n_1; \qquad n_3 = \frac{m}{3}n_2 \ldots \text{etc.}$$

Now, the method of intracellular recording which I have described so far, does not enable us to resolve e.p.p.'s into quantal components *directly*. We only register their summed amplitudes, and as there is an appreciable statistical variation among the sizes of the individual unit potentials, we must choose an indirect analytical method if we want to resolve our observations further. The raw data provided by these experiments are *varying amplitudes of e.p.p.'s*, and we can examine the amplitude histogram and see whether the distribution of sizes fits our quantal hypothesis.

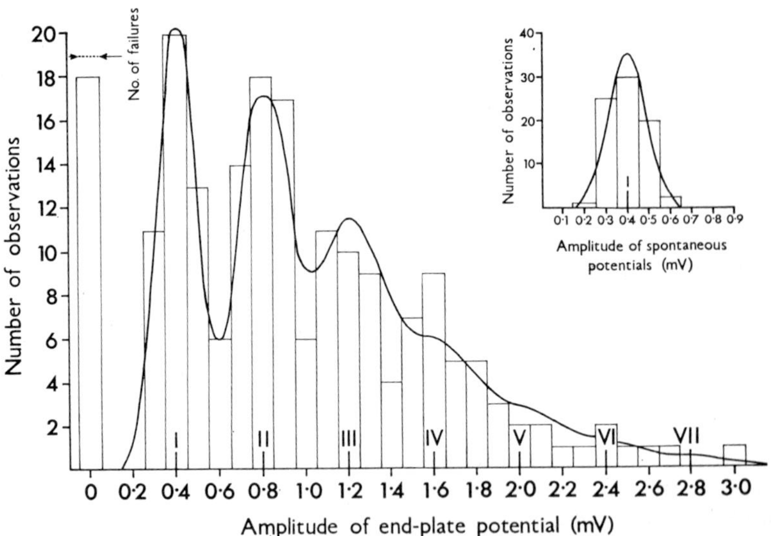

Fig. 10. The distribution of amplitudes of evoked e.p.p.'s, and of spontaneous min. e.p.p.'s (inset), from a mammalian end-plate blocked with high magnesium. (From Boyd and Martin, 1956b.) Peaks of the e.p.p. amplitude histogram occur at 1, 2, 3 and 4 times the mean amplitude of the spontaneous potentials. A Gaussian curve is fitted to the latter and used to calculate the theoretical distribution of e.p.p. amplitudes (continuous curve). Arrows indicate expected number of failures.

An example is shown in Fig. 10 taken from an experiment by Boyd and Martin (1956b) on a cat's tenuissimus muscle. The value of m, determined from equation (2), was 2·33 in this case. Out of 198 impulses, 18 failed to evoke an e.p.p., while the predicted number is 19. The e.p.p. sizes show several peaks; the first peak corresponds closely to the mean amplitude of the spontaneous potentials, the second and third peaks correspond to twice and three times this value. The smooth, multi-modal curve in Fig. 10 is a theoretical amplitude distribution. It was constructed by superimposing a number of bell-shaped curves whose mean values and variances were made equal to 1, 2, 3 etc., times the mean and variance of the spontaneous potentials (Fig. 11). The area under each curve was determined by the number derived from Poisson's law. I have described this somewhat tedious exercise because it shows beyond any doubt that the end-plate potential, under these experimental conditions, is made up of quantal components which are identical, in mean size and in variance, with the spontaneously occurring miniature potential.

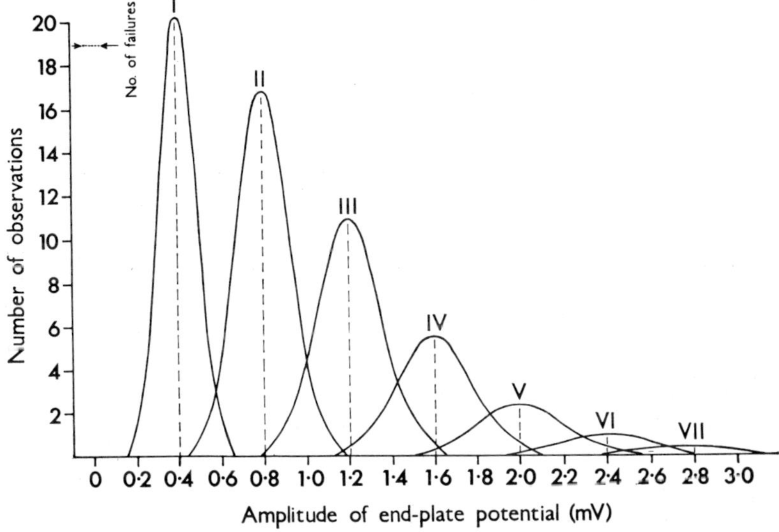

Fig. 11. Method of obtaining the continuous theoretical curve in Fig. 10 (from Boyd and Martin, 1956b). The area under each Gaussian curve was determined by Poisson's theorem, the mean and variance of each curve by integral multiples of the mean and variance of the spontaneous potentials. Algebraic summation of ordinates produces the continuous multi-modal curve of Fig. 10.

RECORDING FROM SMALL PARTS OF A MYONEURAL JUNCTION

The experiments which I have described so far were all made with an intracellular electrode which enables one to record the postsynaptic activity from a single neuromuscular junction. It now became necessary to take our physiological dissection one step further and examine the events which occur within a small part of a single junction. The problems which concerned us were the following: non-myelinated terminal branches of the axon extend over approximately 100μ or even longer distances; what happens when the nerve impulse arrives at the end of the most distal myelin segment, and at the beginning of the synaptic arborization? Does the impulse actively propagate into the terminals? Is there a definite latency between the action potential wave in the nerve ending and the commencement of postsynaptic current flow?

To answer such questions intracellular recording is not good enough, for it does not discriminate between local electrical changes which occur at different parts of the myoneural junction. Because of the high impedance of the cell membrane, the synaptic potential changes spread without much attenuation over an appreciable part of a millimetre along the muscle fibre, and therefore the local membrane effects produced at different points of the myoneural junction are all added and recorded without spatial discrimination. This is demonstrated by the fact that simultaneous records of spontaneous min. e.p.p.'s, made with two intracellular electrodes in the junctional region about 0·5 mm apart, are practically indistinguishable (del Castillo and Katz, 1956b). And yet, it is quite clear that the miniature potentials arise from highly localized events which take place at random spots, distributed all along the myoneural contact, but each confined to a very small part of the junction.

This was shown by withdrawing the micro-electrode from the interior of the cell and probing along the outside of the muscle fibre. With an extracellular electrode, one records in general no potential changes from a resting muscle because the outside fluid bath acts as an effective short-circuit. But if the electrode tip approaches a motor nerve terminal very closely, within a few microns, then one suddenly sees spontaneous miniature potentials (Fatt and Katz, 1952; del Castillo and Katz, 1956; Liley, 1956a; Hubbard and Schmidt, 1963; Katz and Miledi, 1965a). These potentials arise from the convergence of local current into an activated spot of the muscle membrane. In the immediate neighbourhood of this active focus, the current density is so high that an appreciable electric field is set up locally in the external medium. Examples are illustrated in Fig. 12. In each pair of traces, the upper one was recorded with an intracellular, the lower trace was obtained simultaneously with a focally placed external, micro-electrode.

There are several characteristic differences between internal and external miniature potentials. The external potentials are of opposite sign and have a much more rapid time course than the potential change

20

recorded across the fibre membrane. This is because with our focal electrode we are recording the potential drop in the outside bath due to current which converges into the fibre at that point, while the internal electrode registers the resulting depolarization of the fibre membrane, i.e., the capacitatively delayed rise of internal potential. The most important difference, however, between extracellular focal and intracellular potentials is that the former appear much less frequently. Of every 100 internal miniature potentials, only a few are picked up by the external micro-electrode. The reason, of course, is that most of the spontaneous potentials originate at points which are more than 5 to 10μ

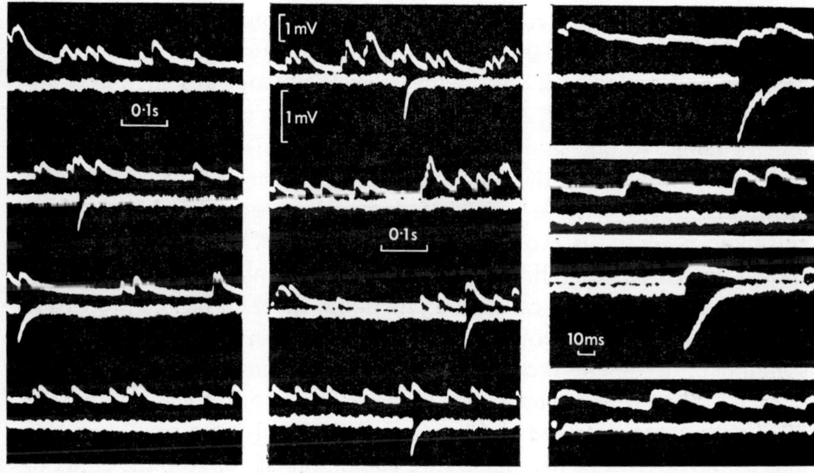

Fig. 12. Simultaneous intra- and focal extra-cellular recording of spontaneous min. e.p.p.'s. (From del Castillo and Katz, 1956b.) In each pair of traces the upper record shows 'internal', the lower the much less frequent 'external' discharges. Note: faster time base in right hand block.

away and consequently outside the range of the external electrode. Thus, we have here a method of recording which has high spatial discrimination; it records selectively from a very small portion of the myoneural junction—about 10μ in length—and it enables us to distinguish between electric events which occur at two points 10 microns apart along a single terminal branch. Incidentally, by carefully mapping out the spots (or rather the 'lines') along the surface of a muscle fibre, from which focal external miniature potentials can be recorded, one traces in effect, the position of non-myelinated axon terminals. Whenever this map was checked after the experiment by the 'cholinesterase-staining' technique, good agreement was found, and the results have left us in no doubt that practically the whole length of the non-myelinated arborization functions as a 'synaptic ending' and is capable of secreting packets of ACh and producing focal min. e.p.p.'s.

The use of the focal calcium pipette

The method of locating active spots on the surface of the muscle fibre is, of course, not very easy, and once the electrode is in a critical focal position, one does not want it to be accidentally dislodged. For this reason alone, it would be impracticable to work on a twitching muscle, and steps must be taken to avoid mechanical response to nerve impulses.

R. Miledi and I did this by eliminating calcium from the external medium and replacing it with magnesium (Katz and Miledi, 1965a, c). Under these conditions, spontaneous min. e.p.p.'s still occur without much change, so our vital landmarks still exist. Also, the nerve impulse still travels down into the intramuscular branches and reaches all the terminals. But, the nerve impulse fails to release acetylcholine, and no end-plate potential is evoked on the other side of the junction. If one uses a micro-pipette filled with a solution of potassium or sodium chloride and places it focally on a junctional spot, all one can record is the occasional spontaneous miniature potential and, when the nerve is stimulated, a small focal action potential from the terminal axon branch. What Miledi and I did, however, was to use a micro-pipette filled with a concentrated $CaCl_2$ solution. By applying a small negative bias to the interior of the pipette, we could regulate or stop the efflux of calcium ions from the tip. If one reduces the negative bias and thereby allows a small quantity of calcium to escape from the pipette, the nerve impulse immediately releases ACh at that spot and evokes a focal end-plate potential. Thus we have a muscle in which the nerve impulses reach all the endings, but fail to secrete transmitter—except in the very small region from which we are recording, and only if we allow calcium to diffuse from the recording electrode.

The procedure is illustrated in Fig. 13. The calcium pipette was placed on a junctional spot, at a distance of 75μ from the end of the last myelin segment of the motor axon. Each record on the left is made of six superimposed traces. In A, a sufficient negative bias was applied to stop effectively the efflux of calcium. Each stimulus is followed after a constant delay by a small spike, due to the arrival of the impulse in the non-myelinated terminal. There is no end-plate response. In B, the negative bias was slightly reduced allowing some calcium to escape.

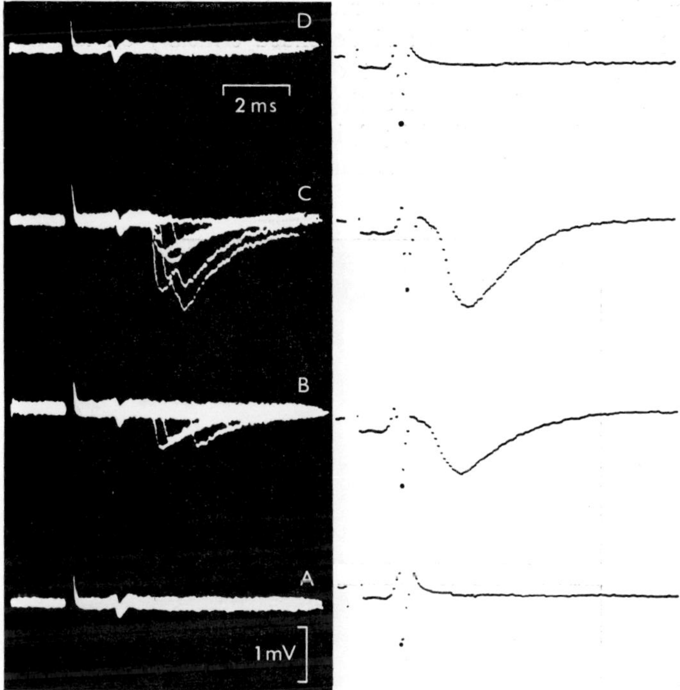

Fig. 13. The use of a 'calcium pipette' in exploring neuromuscular transmission. (From Katz and Miledi, 1965c.) The frog sartorius was immersed in Ca-free solution containing 0·84 mM Mg. A pipette filled with 0·5 M CaCl$_2$ was used to record focal external potentials from a junctional spot. Efflux of Ca was controlled electrophoretically. In the 4 records, A to D, Ca-efflux was stopped initially by applying sufficient negative voltage to the pipette; the bias was then reduced in two steps, and finally re-applied. On the left, several superimposed traces are shown at each stage; on the right, automatically obtained average of 600 traces at each stage. Negative peaks of averaged nerve spikes have been retrouched. Calibrations apply to records on the left.

Now, out of six impulses, three fail to release ACh, while the other three release single packets producing a postsynaptic potential change after a variable delay. In C, by a slight adjustment of the electric bias, more calcium was allowed to accumulate around the nerve ending and this time there was only one failure during six impulses, the other five releasing one or more packets. It should be noted that both presynaptic

and postsynaptic responses were recorded from a small, sharply localized, part of the junction, and depended critically on the electrode
position. A sideways movement of the pipette by several microns
across the fibre surface caused both potentials to vanish. We are recording here genuine 'synaptic delays', i.e., the latent period *at a given spot*
between the flow of membrane currents in the presynaptic and postsynaptic cells.

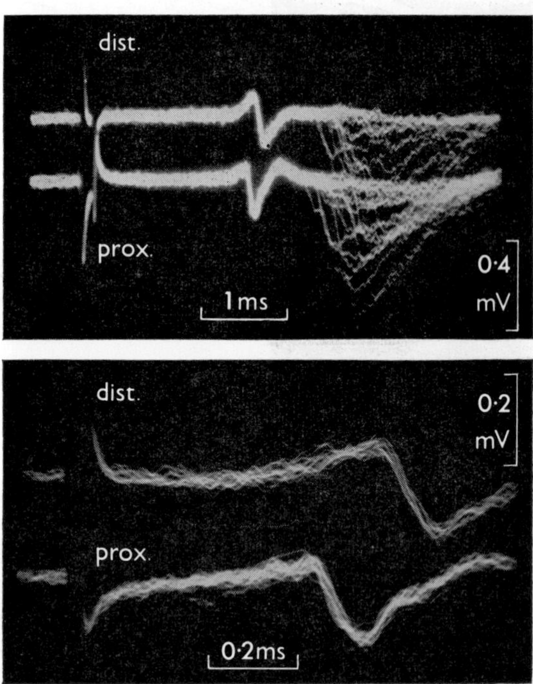

Fig. 14. Extracellular records of focal nerve spikes and
e.p.p.'s from two spots, 60μ apart along a nerve terminal
(the proximal position being 42μ distant from the last
myelin segment). Lower part shows the nerve spikes
alone on a fast time base. (From Katz and Miledi,
1965a.)

If we move the focal electrode to different points along a terminal
axon branch, we obtain similar results, but the potential changes are
shifted slightly along the time base. In Fig. 14 two electrodes have been
placed 60μ apart along a nerve terminal, and we see that the response at
the more distant point is delayed by about an extra 0·15 msec. From this

we obtain a conduction velocity in the non-myelinated terminal of approximately 40 cm/sec (at 24°C) (see also Braun and Schmidt, 1966). This is very much lower than the speed of propagation in the myelinated parent axon (about 30 m/sec)—a difference to be expected, both because of the greatly reduced diameter (only about $1\cdot5\mu$) and the lack of a myelin sheath in the nerve ending.

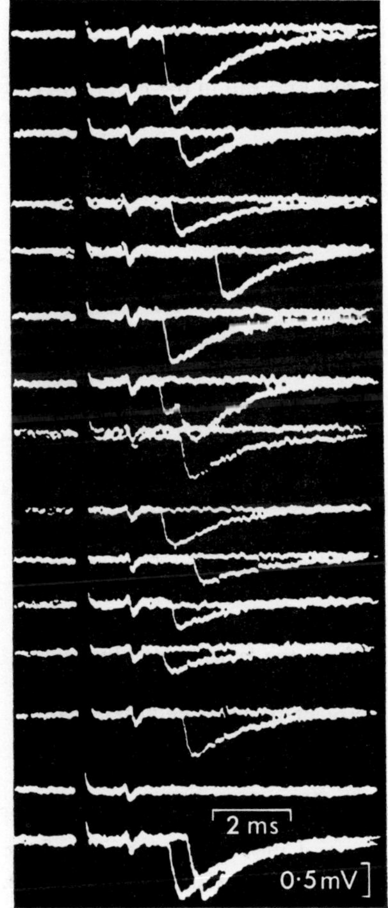

Fig. 15. Focal extracellular records of synaptic response from neuromuscular junction of frog sartorius.
Temperature 17·5°C. Two traces are superimposed in each record. Bias on Ca pipette was adjusted to give approx. 50% failures of postsynaptic response. (From Katz and Miledi, 1965b.)

Synaptic delay and time course of transmitter release

I want to concentrate now a little more on the events which take place at a single recording spot of the junction. Fig. 15 gives us an idea of the statistical variability in successive responses. The presynaptic focal spike is always there, and always at the same time. The calcium level had been adjusted so that there were approximately 50 per cent failures and 50 per cent end-plate responses. Each record is made of two super-imposed traces which show the quantal 'all-or-none' character of the unitary release process. They also show a substantial range of fluctuations in the synaptic delays of different unit responses. A miniature potential may be evoked as early as 0·5 msec, or as late as 2·5 msec, after the arrival of the nerve spike at this point of the junction. Clearly what happens is that the nerve impulse starts up a process which leads, after a certain minimum delay, to a transient rise and fall in the *probability* of quantal release. During this period, packets of ACh have a much greater

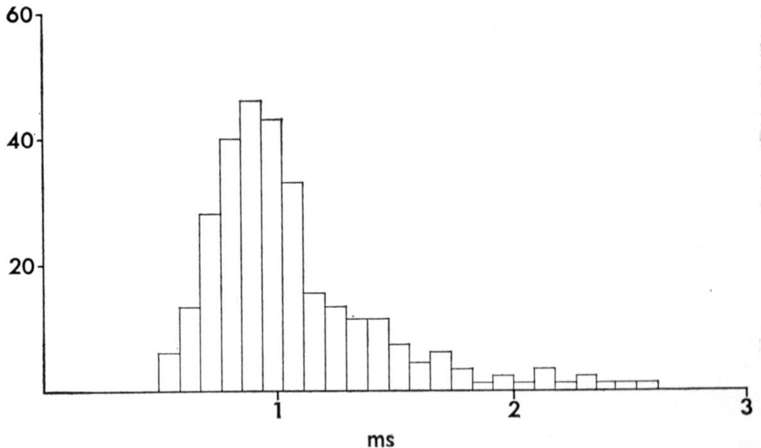

Fig. 16. Distribution of 'synaptic delays' from the experiment illustrated in Fig. 15. (Katz and Miledi, 1965b.) Abscissae: time interval between negative peak of focal nerve spike and commencement of focal end-plate current. Ordinates: number of responses.

statistical chance of being released than during the resting condition. If we make a large series of observations and plot the histogram of the synaptic delays, this will give us the probability distribution in time, that is the time course of rise and fall of the transmitter release. Fig. 16 shows such a histogram. There is a minimum delay of about 0·5 msec after the peak of the presynaptic membrane current before the first postsynaptic unit appears. This is followed by a rapid rise and a slower fall occupying about 2 milliseconds.

It is clear that the secretion of the transmitter is not synchronous with the arrival of the action potential in the nerve terminal, but lags wel behind it, and there appears to be a definite latency before even the first packet makes its appearance. This delay becomes much more conspicuous if one lowers the temperature. In Fig. 17 three examples of the synaptic latency distribution are shown, at 17·5°, 7° and 2·5°C respectively. The histogram of Fig. 16. appears as a brief spike on the slower time scale chosen for Fig. 17. At 2·5°C, the minimum synaptic delay is about 5 msec; the majority of the units are released around 10 msec after the local arrival of the nerve impulse, and the total period of evoked release spreads over more than 30 msec.

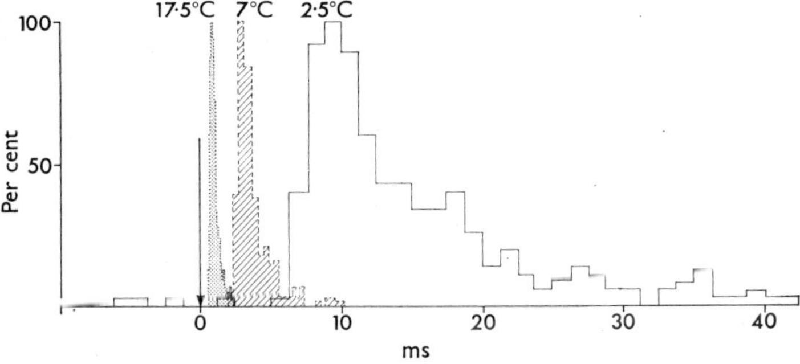

Fig. 17. Histograms of synaptic unit delays at 17·5°, 7° and 2·5°C. (Katz and Miledi, 1965e). Abscissae: as in Fig. 16. Ordinates: percent of maximum (the total numbers of unit responses represented in the three histograms are 292,271 and 265, respectively).

Now, the rise of the postsynaptic current on which all these measurements depend, does not actually signal the moment of release of the corresponding quantal packets of transmitter. After its release from the terminal, the transmitter must diffuse across the synaptic cleft and react with postsynaptic receptors before an electric change takes place. We have, however, convincing evidence that neither diffusion nor reaction times are involved to a significant extent. With ionophoretic pulse application of ACh, depolarization of the junctional region of the muscle fibre can be observed as early as 0·1 to 0·2 msec after the start of the pulse. Under natural conditions when ACh is secreted by the nerve terminal itself, the delay due to local diffusion is presumably much shorter than this. It is clear that the major part of the observed synaptic latency must reside in the mechanism by which the action potential raises the probability of quantal release.

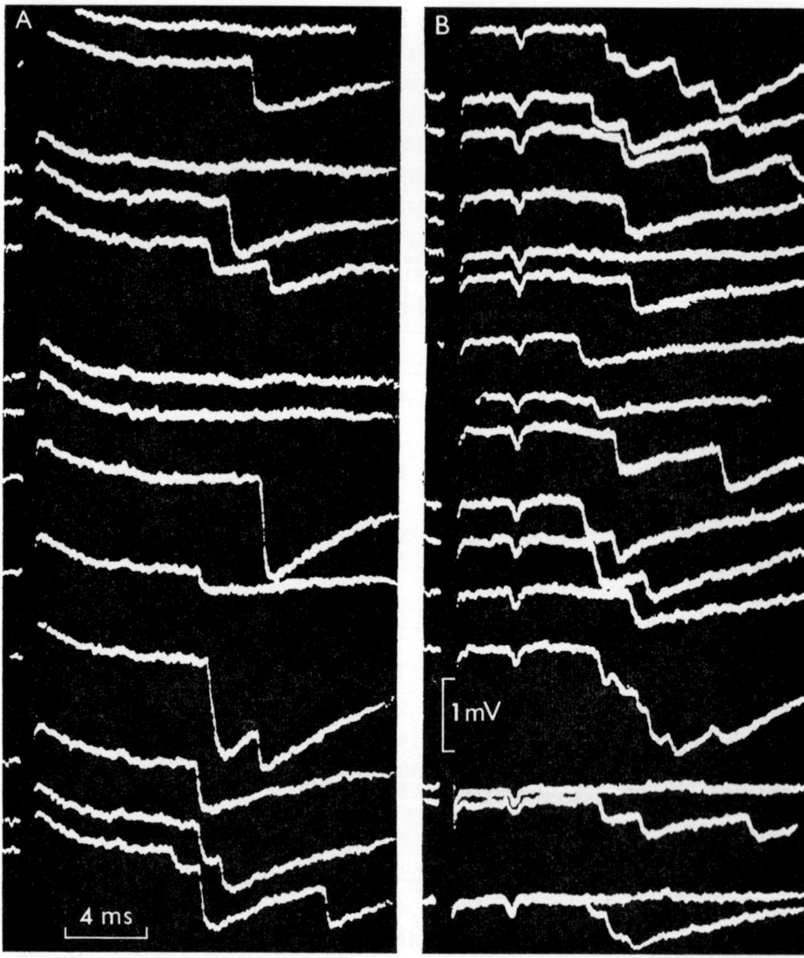

Fig. 18. Focal recording of end-plate response at low temperature. (Katz and Miledi, 1965e). A and B: from two junctional spots. The terminal nerve spike is visible only in B. Note the temporal dispersion of postsynaptic units.

More about Poisson distributions

At low temperature the release of units of transmitter becomes dispersed in time to such an extent that one can count individual packets directly with our method of focal recording, and one does not require the more complicated analysis which I described previously (Fig. 10). Fig. 18 shows the kind of records one obtains at a temperature of about 7°C,

28

with our calcium bias adjusted so that only a small number of packets are being released by each impulse. The presynaptic nerve spike, if it is large enough to be seen, appears at the same moment in every trace; sometimes it fails to evoke an e.p.p., in other traces one sees a variable number of discrete unit potentials appear after variable delays. It is not difficult to resolve individual units and count them, as their numbers are sufficiently small, and their rising phases sufficiently brief to separate one from the other.

This enables us to test the applicability of Poisson's law in a simpler and more direct way than I described earlier on. Table 1 shows the results of an experiment in which several series of observations were made on a junctional region, without moving the recording electrode but with small readjustments of the electric bias on the calcium pipette from one series to the next.

Table 1. The Poisson distribution of quantal end-plate responses (From Katz and Miledi, 1965e)

Several series of responses were obtained from an end-plate, with different settings of calcium efflux from the recording electrode. $n_{0,1,2,\ldots x}$: numbers of observed traces showing $0, 1, 2, \ldots x$ post-synaptic unit responses. Calculated numbers: obtained from Poisson's law $p_x = \dfrac{n_x}{N} = \dfrac{e^{-m}m^x}{x!}$. N = total number of nerve impulses (= total number of traces).

$$m = \text{(total number of units)}/N = (n_1 + 2n_2 + 3n_3 + \ldots xn_x)/N.$$

A.

	n_0	n_1	n_2	n_3	$\geqslant n_4$	Total $N(= \sum n_x)$	Total number of units $(= \sum (xn_x))$
Observed	440	195	37	2	0	674	275
Calc. ($m = 0{\cdot}41$)	448	183	37	5	1		

B.

	n_0	n_1	n_2	n_3	$\geqslant n_4$	Total N	Total number of units
Observed	71	57	13	6	0	147	101
Calc. ($m = 0{\cdot}69$)	74	51	17	4	1		

C.	n_0	n_1	n_2	n_3	$\geqslant n_4$	Total N	Total number of units
Observed	64	49	27	8	0	148	127
Calc. ($m = 0\cdot86$)	63	54	23	7	1		

D.	n_0	n_1	n_2	n_3	n_4	$\geqslant n_5$	Total N	Total number of units
Observed	40	63	45	20	4	1*	173	235
Calc. ($m = 1\cdot36$)	44	60	41	19	6	2		

* This single record showed 6 unit responses.

In series A, a total of 674 nerve impulses were elicited. The electric 'brake' on the calcium efflux had been set so that there were only a few end-plate responses: in fact, on our 674 traces we counted 440 failures, 195 single units, 37 'twins', 2 'triplets' and nothing higher than that: a total of $195 + 74 + 6 = 275$ units, and a mean (m) of $275/674 = 0\cdot41$ packets per impulse. If we use this value of m in our Poisson equation the predicted numbers are those shown in the lower row: 448 failures, 183 singles, 37 twins, 5 triplets and 1 quadruplet. We then readjusted the D.C. bias on the pipette to allow more calcium to escape and repeated the experiment. In series D, for instance, a total of 173 nerve impulses produced $0 \times 40 + 1 \times 63 + 2 \times 45 + 3 \times 20 + 4 \times 4 + 6 \times 1 = 235$ units, so that m was now $235/173 = 1\cdot36$. The predictions of Poisson's theorem are 44 failures, 60 singles, 41 twins, 19 triplets, 6 quadruplets, and 2 quintuplets, as shown in the lower row. The agreement is very satisfactory, though perhaps not quite as good as in the famous textbook example, which shows that during the nineteenth century the annual distribution of fatalities due to horse-kicks in the Prussian cavalry was fitted with great precision (as one would expect) by Poisson's law.

JUNCTIONAL TRANSMISSION WITHOUT IMPULSES

At this juncture R. Miledi and I were concerned to find out more about the nature of the synaptic delay. What are the intermediate steps by which the nerve impulse evokes secretion from the terminal twigs of the axon? We know that the action potential is associated with a sequence of important changes in the axon membrane, first an increase of sodium conductance which leads to an intense influx of Na ions, then a rapid suppression of the sodium permeability and an increase of potassium conductance. These features are essential for the rapid rise and fall of the action potential wave (Hogdkin, 1964). Are they also prerequisites for the subsequent transmitter release? The possible involvement of sodium entry and of the underlying Na conductance change was tested by applying a specific inhibitor of this process, viz., the Japanese puffer fish poison, tetrodotoxin. This substance has an extremely potent 'local anaesthetic' effect and blocks nerve impulses in a concentration about 100,000 times lower than procaine (see Mosher, Fuhrman, Buchwald and Fischer, 1964). Its mechanism has been studied by Narahashi, Moore and Scott (1964) and by Kao (1966) who found that it specifically interferes with the transient rise of sodium permeability which normally occurs when the membrane is depolarized. We have used concentrations of tetrodotoxin of about 10^{-6}g/ml which is more than 100 times the dose needed to block the impulse and which renders both nerve and muscle membranes completely inexcitable by electric stimuli, in the ordinary sense. But as Elmqvist and Feldman (1965) showed, this does not abolish the spontaneous min. e.p.p.'s. And it had previously been found by Furukawa *et al.* (1959) that the sensitivity of the end-plate to applied ACh was unaffected, provided of course one measures the locally produced depolarization of the muscle membrane (and does not look for propagated action potentials or twitches). R. Miledi and I (1965d; 1967a, b) confirmed Elmqvist and Feldman's observation and found that there is no significant change in the frequency or size of the miniature potentials. Moreover, when we used a method of locally depolarizing the motor nerve ending, we were able to increase the probability of quantal release (Figs. 19–21) and to evoke large e.p.p.'s.

This was an interesting result and had several important implications. In the first place, it presented us with a 'pharmacological isolation' of the neuromuscular junction which had not previously been achieved. Other paralytic agents, like curare, or magnesium, or procaine, have a strong blocking action on one of the junctional links, on the postsynaptic receptors or the presynaptic release—either exclusively, or in addition to blocking nerve and muscle conduction. The selective effect of tetrodotoxin, sparing the myoneural junction, while paralysing nerve and muscle fibres on either side, is one more demonstration of the fact that impulse conduction and synaptic transfer from nerve to muscle depend on different mechanisms. The specific pharmacological properties of the myoneural junction were, of course, known since the days of Claude

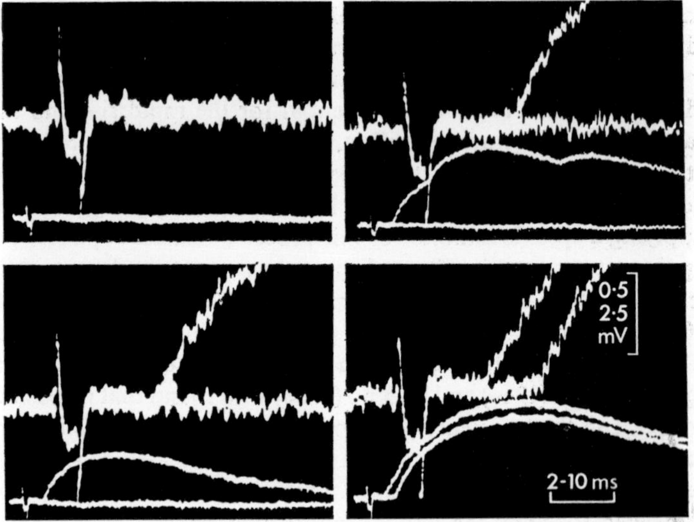

Fig. 19. Quantal e.p.p.'s evoked by brief depolarizing pulses in tetrodotoxin treated muscle. (From Katz and Miledi, 1967b.) Each block shows two simultaneous recordings, the top record at the higher speed and amplification indicated on the calibration scales. Each recording consists of two superimposed traces. The records on the left were obtained with a slightly weaker pulse than those on the right. Note discrete units of response appearing after variable latencies.

Bernard. They are generally interpreted as in keeping with the view that specific acetylcholine receptors are localized at the junctional region of the muscle fibre. But in the past, some authors (e.g., Nachmansohn, 1959) have claimed that the chemical specificity of the myoneural junction is only apparent and arises from the presence of differential diffusion barriers which prevent substances like ACh, curare, or neostigmine, from gaining access to specific sites on the cell membranes except in the region of the myoneural junction. Perhaps we shall now be told that there is another barrier, of precisely converse properties, which prevents tetrodotoxin from reaching the neuromuscular junction while admitting it to the nerve and muscle fibre on either side. Frankly, I see no basis for either of these postulates. The idea of a 'differential barrier' was introduced solely for the purpose of explaining away the specific properties of the neuromuscular junction; there was no positive evidence to support it at any time, and the tetrodotoxin experiments have made that suggestion even more unattractive.

The role of calcium in junctional transmission

The tetrodotoxin experiment also rules out effectively the suggestion that sodium entry into the axon can be an important factor in coupling membrane depolarization to transmitter release. To escape this argument one would have to assert that the motor nerve-endings remain electrically excitable and continue to give an action potential when impulse conduction has been eliminated elsewhere along the axon. But the depolarizing pulses applied to the terminal, which normally evoke a nerve impulse, now produce a localized graded increase in the rate of transmitter release. There is as a rule no sign of a critical theshold or of a regenerative response (for exceptions, see Katz and Miledi, 1967b, p. 35). This indicates that Na-entry into the nerve terminal has been either eliminated or reduced to a very low level at which electric excitation is no longer supported.

The ion which *is* indispensable for the electrically evoked transmitter release is calcium. This is true normally, as well as in the tetrodotoxin-treated preparation. Fig. 20 illustrates this point. The muscle has been paralysed with the toxin and placed in a bath containing magnesium instead of calcium. In all records (A–E) a depolarizing pulse was applied to the nerve terminal, but failed to evoke an e.p.p. except in B and D where an ionophoretic dose of calcium was added beforehand. This is shown again in record F, taken on a slow time base: calcium was allowed to diffuse from a separate micropipette during the period between the arrows. This promptly enabled the depolarizing pulses to become effective and evoke a postsynaptic response.

The next question which R. Miledi and I took up concerns the stage of the presynaptic process at which external calcium ions are utilized. Is it necessary for calcium to be available on the outside of the membrane during the depolarization, or will it suffice to apply calcium afterwards, at a time when the probability of release normally rises to its peak?

We have tested this question by using focal twin-pipettes. One barrel was filled with a NaCl solution and served as a depolarizing electrode; the companion barrel contained $CaCl_2$ and was used to raise the local calcium concentration at desired moments, before or after the depolarizing pulse. The answer was quite clear (Katz and Miledi, 1967c): calcium must be present during the depolarization; if one delays the calcium pulse until the end of the depolarization, no release occurs.

I should like now to reconsider our working hypothesis and perhaps fill in some additional detail. I suggested that depolarization somehow changes the properties of the terminal axon membrane so that it pre-

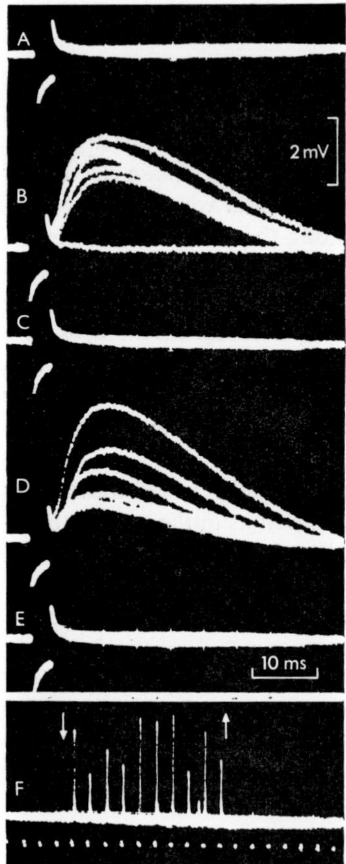

Fig. 20. Effect of calcium in tetrodotoxin paralysed muscle. (From Katz and Miledi, 1967a.) The calcium content of the Ringer had been replaced by 1·7 mM magnesium. Focal application of depolarizing pulses at 2 per sec. A separate pipette containing 1 M $CaCl_2$ was placed nearby, and Ca was released ionophoretically during B and D, and during the period between arrows in the low-speed record F. in A, C and E, when calcium was withheld, no e.p.p. was evoked by the pulses (six superimposed traces in each record). In F, dots indicate depolarizing pulses, at 0·5 sec intervals.

sents many more reactive sites to the colliding vesicles and thereby raises the statistical chances of quantal release. R. Miledi and I now put forward a more specific suggestion: namely that depolarization opens a gate to calcium ions (or—what amounts to the same thing— that depolarization makes available specific carriers for calcium ions in the membrane). As a consequence of this increased 'calcium conductance', Ca ions can move—down a very high concentration gradient— towards the inside of the axon membrane and thus reach the critical sites of the release reaction. We are suggesting that, at these sites, calcium is essential for the process which causes a transient fusion of axon and vesicular membranes and which leads to the release of a quantal packet of transmitter.

This hypothesis has an interesting consequence. As we go on depolarizing the axon terminal, i.e., raising its internal potential, two things will happen. (1) We increase the calcium conductance. This is a change which may be expected to occur with a small delay; its rise and fall probably lag somewhat behind the rise and fall of the depolarization. Moreover, this process presumably reaches maximum intensity at a certain level of membrane potential. (2) The rise of the internal potential must, by Ohm's law, oppose the entry of cations. This is an immediate ionophoretic effect; it has no measurable delay and would go on increasing in proportion to the imposed voltage change without saturation. Thus, if we 'overdo' our depolarization and displace the internal potential to a sufficiently high positive value, we should reach the equilibrium for calcium ions and at that point should stop the entry of calcium into the membrane.

In other words, if we can produce a large enough voltage step across the membrane and maintain it for a long time, say 10 msec or more, we may find that the release of the transmitter is completely suppressed *during* the pulse and only occurs as a surge at the moment when the voltage step terminates.

To test this prediction, the neuromuscular junction is not very suitable because the small size of the nerve terminal prevents a direct measurement of the presynaptic potential change. All we can do here is to apply a rectangular current pulse through a focal micro-electrode and hope that the voltage step imposed on the small area of membrane under the electrode varies in proportion to the applied current intensity. There are reasons for believing that this can be true only for a limited range of intensities and durations. R. Miledi and I were unable to suppress transmitter release for the whole period of a depolarized current if it lasted more than 3 to 4 msec (Katz and Miledi, 1967b). But within that range we found that lengthening the duration of the pulse had two opposing actions: (i) it greatly increased the number of ACh packets which were released; (ii) it *delayed* the release of the first packet until some little time had elapsed after the end of the pulse (see Fig. 21). This peculiar 'latency shift' was difficult to explain except by invoking two mutually opposing effects of the voltage step, and it provided the first indication that our detailed hypothesis has a good deal of validity.

The giant synapse of the squid

R. Miledi and I then decided to look for a preparation which was more suitable for our purpose. The obvious choice was the giant synapse in the stellate ganglion of the squid. Here, the terminal branch of the

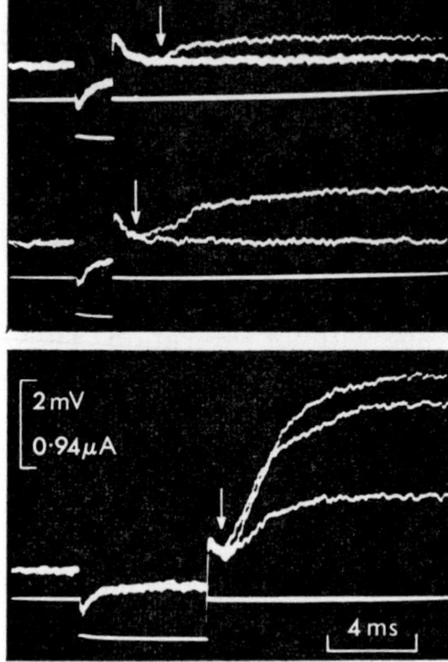

Fig. 21. Example of 'latency shift' of e.p.p. observed when the duration of a depolarizing pulse is lengthened. (From Katz and Miledi, 1967b.) In each record, the upper traces show fluctuating e.p.p.'s evoked by focal current pulses which are monitored in the lower traces. Beginning of e.p.p.'s shown by arrows. Note: lengthening the pulse increases *and delays* the response. Temperature 4·5°C.

presynaptic axon is large enough to accommodate one or two intra-cellular micro-electrodes (Bullock and Hagiwara, 1957; Takeuchi and Takeuchi, 1962). There is very good converging evidence that transmission is brought about by a chemical mediator (Hagiwara and Tasaki, 1958; Miledi and Slater, 1966), but its identity has not yet been revealed. Transmission occurs with a characteristic synaptic delay between the arrival of the impulse in the terminal and the commencement of post-synaptic current flow. The effect of the impulse in evoking a postsynaptic change depends on the presence of calcium and is reduced by adding magnesium to the external solution. The postsynaptic potential can be shown to arise from a local ionic conductance change and to reverse in polarity when the membrane potential of the giant axon is displaced beyond a certain level (about 20 mV *positive* inside, according to unpublished observations by R. Miledi). Furthermore, Miledi (1966, 1967) has recently been able to demonstrate the presence of spontaneous post-synaptic potentials which are analogous in every respect to the min. e.p.p.'s. They are, however, of such small amplitude—a consequence of the low input impedance of the giant axon—that they are very difficult to detect. In spite of the incomplete state of the evidence, we can be certain that the transfer of electric signals across this synapse is effected by the quantal release of a transmitter substance which raises the ion permeability of the postsynaptic membrane and produces a depolarization similar to the end-plate potential.

A further analogy to the neuromuscular junction was found when

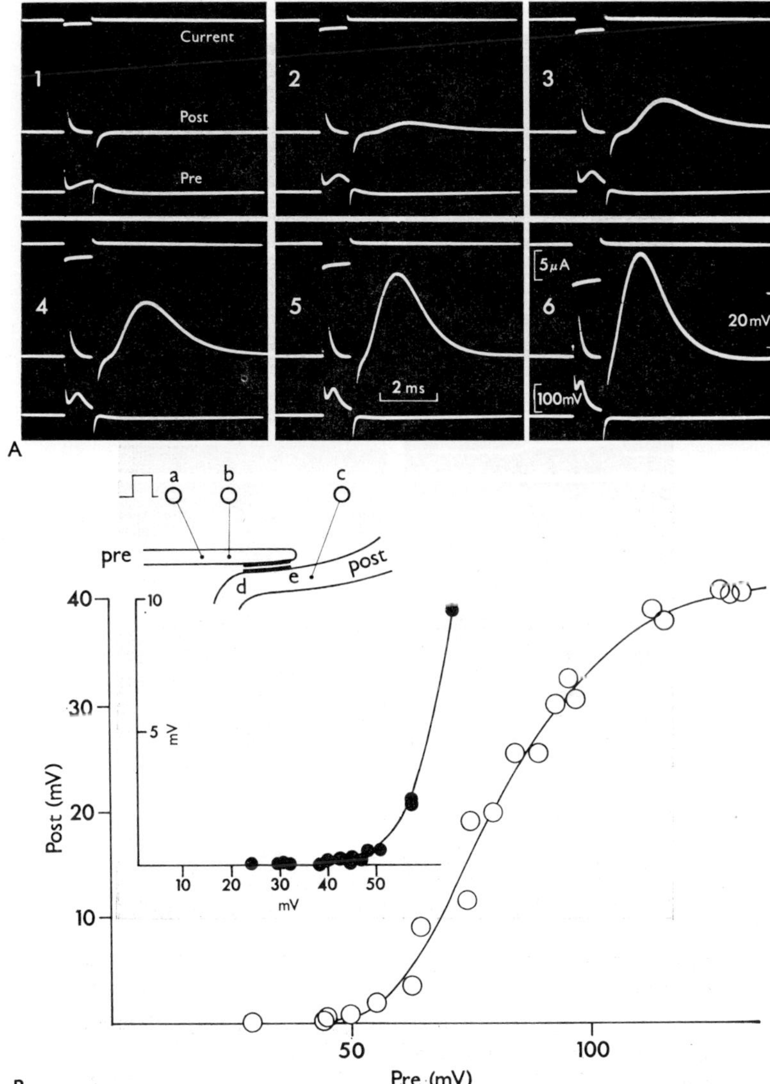

Fig. 22. Input/output relation of squid giant synapse, treated with tetrodotoxin (2×10^{-7}g/ml). (From Katz and Miledi, 1966.) Arrangement of intracellular electrodes shown in upper diagram of B. Pre: presynaptic terminal. Post: postsynaptic giant axon. Length of synaptic contact d–e: 0·8 mm. a: Current-passing electrode; b: 'pre'-recording electrode; c: 'post'-recording electrode. Distances: a–d, 0·7 mm; b–d, 0·35 mm; e–c, 0·3 mm. A: Sample recordings. B: Input-output relation obtained with 1 msec current pulses. Abscissae, presynaptic depolarization; ordinates, postsynaptic response. Inset, initial part of curve in greater detail.

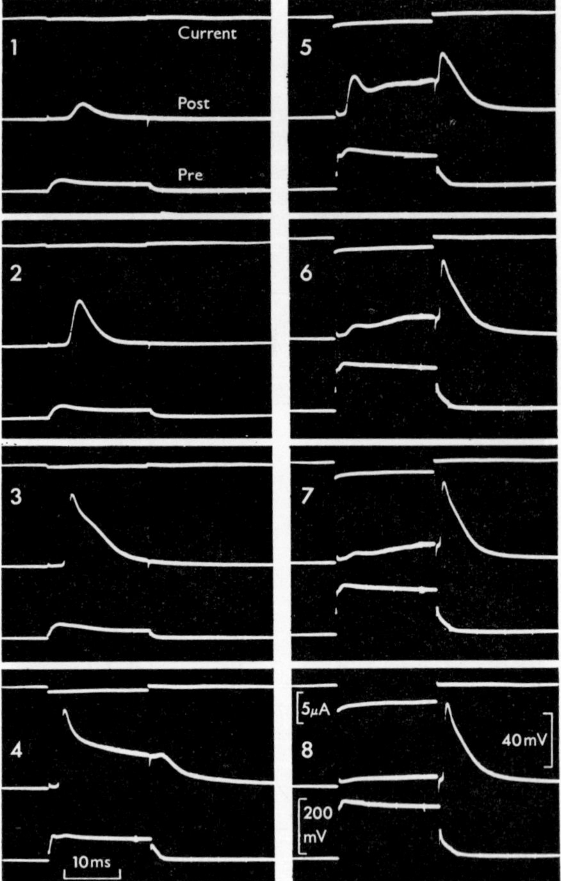

Fig. 23. Suppression of transmitter release during large 'positive voltage step' of presynaptic membrane potential. (From Katz and Miledi, 1967d.) Squid giant synapse treated with tetrodotoxin. Presynaptic terminal loaded with tetra-ethylammonium ions. Blocks 1 to 8: increasing pulse intensity. In each block of records, the bottom trace shows the presynaptic voltage step, the middle record shows potsynaptic response, while the current pulse is monitored in the top trace. Note progressive suppression of 'on'-response, and replacement by 'off'-response, as presynaptic voltage is increased from 100 to about 200 mV (records 4 to 8).

tetrodotoxin was used. After nerve impulses had been eliminated by the toxin, local depolarization of the nerve terminal still evoked a post-synaptic potential change, provided calcium was present in the bath (Bloedel *et al.*, 1966; Katz and Miledi, 1966, 1967d; Kusano *et al.*, 1967).

Experiments on this synapse have provided information which is complementary in certain respects to that obtained from the motor end-plate. By direct measurements of both, presynaptic and postsynaptic potentials, the whole input/output relation of this synapse can be followed quantitatively. Miledi and I found, for instance, that brief depolarizing pulses of 30 mV, or more, elicit postsynaptic responses which occur after a definite local delay (Fig. 22), and that the initial part of the synaptic transfer characteristic is very steep, the response increasing about ten-fold for an additional depolarization of 10 mV.

To produce a large voltage step and maintain the inside of the axon terminal at a high positive potential, it is necessary first to reduce the potassium conductance of the membrane. This can be achieved (see Hagiwara and Tasaki, 1957; Armstrong and Binstock, 1965) by iono-phoretic injection of tetraethylammonium into the terminal axoplasm. After these preliminaries we are in a position to do our test experiment, and the results are illustrated in Fig. 23. It can be seen that with increasing positive-going voltage steps, a point is reached beyond which the synaptic transfer *during* the current pulse is diminished and finally, with a voltage step of approximately 200 mV, nearly completely suppressed. At the end of the pulse, as soon as the positive internal potential of the presynaptic ending drops, a large postsynaptic response suddenly appears. This indicates that the large voltage step which has made the inside of the terminal 130 mV positive with respect to the outside, acts as an electric potential barrier; as soon as the barrier drops, transmitter is liberated in a large surge. The results fit qualitatively very well into our picture, according to which, entry of calcium from the outside into the membrane is a prerequisite for the reaction which leads to the release of transmitter. Various estimates have been made of the concentration gradient of calcium ions across the axon membrane. With 11 mM/1 in the external solution and less than 10^{-5}M in the axoplasm (Hodgkin and Keynes, 1957), the equilibrium potential for calcium ions would be at least 85 mV positive inside. If the internal concentration of ionized calcium is 4×10^{-7} M, no discrepancy between result and hypothesis would arise.

Commentary

I have tried to present a long series of experiments on transmitter release in a coherent fashion without allowing myself to be deflected by too many asides. But now, perhaps, it is not out of place to expand on certain points, even though they are outside the main line of my argument, and some of them surrounded by controversy.

A comment on terminology

The first point concerns terminology, a somewhat tedious matter which I am discussing only with reluctance and after much provocation. Controversies about words, like arguments about priority, are dominated by emotion, and I well remember W. Feldberg's dictum, namely that there is a type of scientist who, if given the choice, would rather use his colleague's toothbrush than his terminology! My colleagues and I were looking for an adjective which would adequately describe the important property of evoked transmitter release, namely, that it occurs in standard 'packets' of large multi-molecular size which are identical with the spontaneously occurring units, and whose size is independent of the event (e.g., impulse, local potential change, chemical or osmotic stimuli) which causes the release. We chose the term 'quantal' for this purpose, which seems entirely proper and unobjectionable to me. I have, nevertheless, found myself challenged on two grounds: (a) for supposedly basking in the reflected glory of quantum physics, and (b) for applying the term 'quantum' to something which is not constant in size, but subject both to random variation and to experimental change. Objection (a) is, of course, impossible to disprove, and to protest would be in vain. All I would say is that I take my authority for the use of words from an ordinary dictionary (the entry 'quantum' in the *Concise Oxford English Dictionary* may serve), and not from books on quantum physics. This may not satisfy the objectors, but I will take that risk rather than discard an adjective which is singularly apt in describing a whole set of characteristic features. Objection (b), namely that the word 'quantum' is not applicable to something of variable size, seems to be based on a misconceived analogy to the physical energy quantum, and I would invite those who raise it to consult a textbook of physics as well as the *Concise*

Oxford. (It is not the energy quantum *hv*, but its coefficient *h* which is a universal constant.)

Variability of the quantal packet

It is quite true that there is considerable variation in size among the individuals which make up the population of quantal packets. Judging from the dispersion in amplitudes of miniature potentials, individual sizes vary over a more than fivefold range. There are many reasons that might account for this degree of scatter. One of the obvious explanations could be a variation in the size of the vesicles which we think contain the packets.

Quantitatively it would be difficult to establish a correlation, because the preparation of very thin sections for the electron microscope is likely to introduce an additional, artificial source of scatter. Measurements which have been made on the rat diaphragm show a threefold range of diameters among synaptic vesicles (Andersson-Cedergren, 1959) which is more than required to explain the dispersion of miniature potentials.

As I have pointed out in an earlier chapter, the mean size and statistical scatter of the miniature units is not altered by those changes in the presynaptic membrane potential which bring about an increase in the frequency amounting to several orders of magnitude. For a long time it seemed that all one could achieve by acute experimental procedures was to alter the statistical probability of occurrence of a quantal discharge, but not the size of the packet itself. Del Castillo and I suggested (1956a) that those agents whose direct effects are confined to the surface membrane are not likely to change the 'quantum size', but that interference with intracellular synthesis of the transmitter might be expected to cause a reduction in the size of the individual packets. Several experimental attempts were made in this direction, but they were unsuccessful until Elmqvist and Quastel (1965) reported that a small dose of hemicholinium which is known to cause depletion of the ACh stores during prolonged nerve stimulation, led to a progressive diminution in the size of the ACh packet. Hemicholinium is believed to stop intracellular synthesis of ACh indirectly, by blocking the uptake of choline from the external medium. The observations of Elmqvist and Quastel suggest that, in the absence of continued synthesis, all the remaining packets diminish in size when some of them are released. This would be expected if there is a continuous dynamic flux of ACh between the vesicles and the surrounding cytoplasm. We have assumed earlier on that vesicles actively accumulate ACh from a low level in the cytoplasm. If this process leads to the building up of a concentration ratio (rather than a fixed final concentration), a continual readjustment and reduction of all the packets would take place when there is continued release without replenishment. In other words, when intracellular synthesis is blocked, the release and loss of the contents of any one packet will then affect the contents of all the other packets; and after a long period of stimulation all the units will be diminished in size to about the same extent.

Quantal and non-quantal release

We concluded from the continuously graded effects of applied ACh that we cannot with our present methods detect the action of individual molecular impacts. This was an important point, because it followed that discrete spontaneous potentials like min. e.p.p.'s cannot be attributed to random leakage of molecules, but must arise from the simultaneous release of a large number of molecules. But by the same argument, we cannot rule out the possibility that, in addition to the intermittent quantal release, diffuse random leakage also occurs. Let us suppose the quantal release makes up only a small fraction of the total loss of ACh from the resting terminal, and there is in addition a steady diffuse leakage of a ten times larger amount (per second). Even at that rate, the steady escape of ACh would produce only an insignificantly small steady depolarization (less than 0·1 mV) of the junctional region of the muscle fibre and so remain undetected.

We can, however, state quite definitely that such a diffuse leakage does not take part in the process of release by an impulse. The proof rests on the successful application of the Poisson analysis (see Fig. 10). This has shown that the end-plate potential evoked by an impulse or by local depolarization of the terminal is purely quantal in nature, and arises from a brief burst of an integral number of ACh packets. If any significant 'non-quantal' component had contributed to the e.p.p., the Poisson theorem would have failed to fit our results.

Is the release of acetylcholine an 'auto-catalytic' process?

The success of the Poisson analysis is important in another respect. It shows that individual *packets* are liberated in a statistically independent random fashion, that is without mutual interference or reinforcement. Some authors hold the view that transmitter release is a self-reinforcing process whereby a small amount of ACh precedes and facilitates the release of a much larger quantity.

I do not see how this idea can be upheld in the case of the neuromuscular junction. It is clear from the statistical behaviour (Fig. 10 and Table 1) that the release of one packet does not raise the probability of release of another packet. If this were the case, the distribution would significantly depart from Poisson expectations and show relatively few 'single' and many more 'multiple' occurrences. Deviations from pure Poisson statistics are, indeed, observed on numerous occasions, but they are of a different kind and usually arise from slow periodical fluctuations or from gradual drift in the probability and mean number m.

One might further argue: is it possible that the process which leads to the release of an individual quantal packet is itself a regenerative event, in which a few molecules of ACh escaping initially lead to the discharge of a few thousand molecules? A counter-argument is that applied doses of ACh, sufficient to produce a small graded depolarization of the muscle membrane, do not evoke min. e.p.p.'s. A steady dose of ACh does not produce a significant increase in the rate of spontaneous potentials, even though there is reason to believe that the ACh causes a small depolarization of the non-myelinated nerve endings. Evidently, if it does so the depolarization is too weak to raise the frequency of miniature potentials noticeably.

Thus, there is no evidence for an obligatory self-reinforcing step in the transmitter release, neither at the uni- nor multi-quantal levels, whether spontaneous or evoked by a nerve impulse.

While there is in general no mutual interaction between quantal events, there are certain observations which might be taken as exceptions to this rule. I have mentioned already the occasional appearance of high-frequency bursts of spontaneous potentials in an otherwise random series of discharges. This phenomenon is still unexplained. Another instance is the occurrence of 'giant' or 'multiquantal' spontaneous potentials described by Liley (1957). These are comparatively rare and inconstant events whose origin is still obscure. The multiquantal discharges might be due to some interaction between the release of one packet and the next (as suggested by Martin and Pilar, 1964), but it is also possible, as suggested by Liley (1957), that infrequent fusions occur among the population of synaptic vesicles and give rise to occasional spontaneous release of 'twins', 'triplets', etc. To a different category belong the large spontaneous e.p.p.'s in guanidine and digoxin treated preparations (Otsuka and Endo, 1960; Birks, 1963) which probably arise from fluctuations of membrane potential and initiation of abortive

44

spikes in the nerve endings. Finally, an apparent exception to the rule of 'non-regenerative release' is the after-discharge in motor axons which originates in anti-esterase treated mammalian muscle.

Repetitive 'back-firing' in mammalian nerve-muscle preparations

This interesting feature was discovered by Masland and Wigton (1940) and Feng and Li (1941). Feng and Li investigated the phenomenon carefully and showed that, while repetitive firing of nerve endings can arise from various other changes in the chemical environment, the effect produced by cholinesterase inhibitors is almost certainly due to local accumulation of ACh around the axon terminals after each impulse.

The repetitive firing of the motor nerve is a rather puzzling phenomenon: it occurs in mammalian, but not in frog, muscle and it apparently originates only at a very small percentage of end-plates in any one muscle. Thus, when Miledi and I tried to elicit the effect by ionophoretic application of edrophonium, we succeeded only at a very small number of the many end-plates we tried. It looks as though the underlying depolarizing action in the nerve endings is weak and generally below threshold, and only at a few end-plates reaches the level required for re-excitation.

It seemed to me until a few years ago, that it was not at all clear whether this re-excitation was brought about by ACh acting directly on the nerve endings from which it was released, or indirectly through the known changes which ACh produces in the adjacent muscle membrane. I suggested that local efflux of potassium from the depolarized area of the muscle fibre might have something to do with it (Katz, 1962). This apparently aroused considerable wrath, and I was told that 'there is overwhelming experimental evidence against this attractive suggestion' (Eccles, 1964, p. 130). However, the combination of (i) a large depolarization (during e.p.p. and muscle spike) and (ii) high local cation permeability maintained by the transmitter, must result in an intense local efflux of potassium from the muscle fibre into the synaptic cleft which might affect the nerve ending, and it seemed to me that the evidence available at the time was not sufficiently relevant to eliminate this possibility.

My suggestion may have provoked the experiments reported by Barstad (1962) and by Randić and Straughan (1964) which make my 'potassium explanation', if not untenable, at least very unlikely. These authors found that one can cut off the muscle fibres on either side of the end-plate, and in this way greatly reduce the postsynaptic membrane potential *without* abolishing the retrograde discharge of motor nerve impulses during esterase inhibition. This kind of experiment strongly indicates that we have to do here with a direct depolarizing action of ACh on the nerve endings.

What seems to me the most probable explanation of the available, still rather incomplete, results is that there are ACh-receptors present in the presynaptic membrane, but at such low density that a release of ACh which almost completely depolarizes the muscle fibre, produces only a few millivolts change in the nerve. In general, the effect is below threshold and too weak to accelerate the rate of ACh release. In the frog this applies to all the myoneural junctions, in mammals to the majority. But at a few endings in mammalian muscle the potential change just reaches the threshold of re-excitation. The impulse coming from one nerve ending then spreads throughout the axon branches and re-excites the whole group of muscle fibres attached to them (cf. Eccles, Katz and Kuffler, 1942). So, the final effect is quite powerful. On the other hand, anything which even slightly raises the threshold of excitation or interferes with ACh action (e.g., a subparalytic dose of curare) will abolish the 'back-firing' altogether. Now, it is on observations of just this kind that Riker and his associates (see Riker, 1966) have based the novel hypothesis that the *principal action* of substances like ACh, curare, antiesterases etc. is presynaptic, and they have stated that there is no good evidence that these substances have a postsynaptic effect. Riker (1966) chooses to disregard the fact that the end-plate sensitivity to ACh is as high after degeneration and disappearance of the nerve terminals as before. He dismisses as irrelevant the electrophysiological measurements of the end-plate activity, of the chemo-sensitivity of the junctional and peri-junctional parts of the muscle fibre, and of the local changes following denervation (Miledi, 1960a, b); he has no use for the detailed studies of min. e.p.p.'s and, indeed, for all the other evidence which has been reviewed here. I agree that it is possible to entertain the hypothesis of Riker if, and only if, one is prepared to ignore everybody else's work on the subject.

Generality of quantal transmitter release

Most of our evidence on quantal release, and on the identity of spontaneous and impulse-evoked 'packets of transmitter' comes from the neuromuscular junction of a few vertebrate species. Over the years, results have accumulated from different sources which indicate that basically the same process goes on at many other peripheral and central synapses where chemical mediators are involved in the transmission of impulses. Min. e.p.p.'s are found at all kinds of vertebrate nerve-muscle junctions, from fish to man (Takeuchi, 1959; Elmqvist, Johns and Thesleff, 1960) and in 'slow' as well as 'fast' muscle fibres (Burke, 1957). They also occur at the myoneural junctions of crustacea (Dudel and Orkand, 1960; Dudel and Kuffler, 1961) and insects (Usherwood, 1961) where the excitatory transmitter substance still remains to be identified (glutamate?—see Takeuchi and Takeuchi, 1964). Quantal release and spontaneous synaptic potentials have been shown in sympathetic ganglia (Nishi and Koketsu, 1960; Blackman, Ginsborg and Ray, 1963 a, b) and at neurones in the stellate ganglion of the squid (Miledi, 1966, 1967). Spontaneous junction potentials occur in smooth muscle (Burnstock and Holman, 1962; Tomita, 1967), and in central neurones of the isolated spinal cord of the frog (Katz and Miledi, 1963). The work of Kuno (1964), Burke and Nelson (1966), Burke (1967) and Eide *et al.* (1967) indicates that unitary synaptic potentials in mammalian spinal neurones can be analysed statistically and shown to be made up of quantal components. There is evidence that inhibitory as well as excitatory effects are transmitted in quantal steps, and spontaneous inhibitory potentials have been observed in arthropod muscle (Reuben and Grundfest, 1960) and at synapses in the spinal cord (Katz and Miledi, 1963).

To summarize, the results from synapses other than skeletal nerve-muscle junctions are somewhat fragmentary, but they suggest that the quantal mechanism of transmitter release is widespread and probably applies to all presynaptic terminals in which chemical transmitters, of different kinds, are synthesized. Moreover, the importance of calcium and the antagonistic role of magnesium in the release process have been verified at several of these different kinds of junctions, and it appears that the control by calcium and magnesium ions is another feature which is characteristic generally of neural transmitter release.

This brings us to the possibility of an even wider generalization: for it appears that the crucial role of calcium is shared by many other types of secretory mechanism in which quantal release of a specific substance is evoked, whether through nerve impulses or through changes in the chemical environment of the cell (see Douglas, 1966; Katz and Miledi, 1967).

Structural and biochemical evidence related to quantal release of acetylcholine

In recent years, there has been remarkable progress in the study of metabolic pathways by which catecholamines are manufactured and stored in adrenergic nerve endings, and in the cytochemical localization of these important neural agents (see von Euler, Rosell and Uvnäs, 1966). By comparison, the biochemical and ultrastructural exploration of cholinergic terminals is still at an early stage. A very significant advance was made by de Robertis and his colleagues (1961) and by Gray and Whittaker (1962) when they showed that it is possible to isolate a particulate fraction from brain homogenates which is rich in nerve endings and contains most of the tissue-bound acetylcholine. These nerve endings had been torn away from their axons, but still retained their usual complement of mitochondria and synaptic vesicles. More recently, the same groups of workers succeeded in disrupting the nerve-ending particles and obtained a fraction of what appear to be synaptic vesicles still loaded with a large part of the tissue acetylcholine. To make all the ACh assayable, the vesicles had to be subjected to further destructive treatment. There are differences of opinion as to whether these vesicles merely accumulate and store ACh, or also contain the synthetizing enzyme system. From the work of Whittaker *et al.* (1964) it appears that this choline acetyltransferase is normally located outside the vesicles, in the terminal axoplasm, but may become adsorbed to the vesicular surface when the particles are suspended in an electrolyte deficient sucrose medium.

There is clearly scope for very important developments along this line of biochemical-plus-ultrastructural approach. One may hope that, by using a cholinergic rather than mixed population of nerve endings, and with further development of the fractionation technique, a quantitative assessment of the average content of the vesicles and of their capacity to accumulate ACh will soon be in sight.

Bibliography

ANDERSSON-CEDERGREN, E. (1959. Ultrastructure of motor end plate and sarcoplasmic components of mouse skeletal muscle fiber. *J. Ultrastructure Res.*, suppl. 1, 191 pp.

ARMSTRONG, C. M., and BINSTOCK, L. 1965. Anomalous rectification in the squid giant axon injected with tetraethylammonium chloride. *J. gen. Physiol.* **48**, 859–872.

BARSTAD, J. A. B. 1962. Presynaptic effect of the neuro-muscular transmitter. *Experientia* **18**, 579–581.

BIRKS, R. I. 1963. The role of sodium ions in the metabolism of acetylcholine. *Canad. J. Biochem. Physiol.* **41**, 2573–2597.

BIRKS, R., HUXLEY, H. E., and KATZ, B. 1960. The fine structure of the neuromuscular junction of the frog. *J. Physiol.* **150**, 134–144.

BIRKS, R., KATZ, B., and MILEDI, R. 1960. Physiological and structural changes at the amphibian myoneural junction, in the course of nerve degeneration. *J. Physiol.* **150**, 145–168.

BLACKMAN, J. G., GINSBORG, B. L., and RAY, C. 1963a. Spontaneous synaptic activity in sympathetic ganglion cells of the frog. *J. Physiol.* **167**, 389–401.

BLACKMAN, J. G., GINSBORG, B. L., and RAY, C. 1963b. On the quantal release of the transmitter at a sympathetic synapse. *J. Physiol.* **167**, 402–415.

BLOEDEL, J., GAGE, P. W., LLINÁS, R., and QUASTEL, D. M. J. 1966. Transmitter release at the squid giant synapse in the presence of tetrodotoxin. *Nature, Lond.* **212**, 49–50.

BOYD, I. A., and MARTIN, A. R. 1956a. Spontaneous subthreshold activity at mammalian neuromuscular junctions. *J. Physiol.*, **132**, 61–73.

BOYD, I. A., and MARTIN, A. R. 1956b. The end-plate potential in mammalian muscle. *J. Physiol.* **132**, 74–91.

BRAUN, M. and SCHMIDT, R. F. 1966. Potential changes recorded from the frog motor nerve terminal during its activation. *Pflügers Arch. ges. Physiol.* **287**, 56–80.

BROOKS, V. B. 1956. An intracellular study of the action of repetitive nerve volleys and of botulinum toxin on miniature end-plate potentials. *J. Physiol.* **134**, 264–277.

BROWN, G. L., DALE, H. H. and FELDBERG, W. 1936. Reactions of the normal mammalian muscle to acetylcholine and eserine. *J. Physiol.* **87**, 394–424.

BULLOCK, T. H. and HAGIWARA, S. 1957. Intracellular recording from the giant synapse of the squid. *J. gen. Physiol.* **40**, 565–577.

BURGEN, A. S. V., DICKENS, F. and ZATMAN, L. J. 1949. The action of botulinum toxin on the neuro-muscular junction. *J. Physiol.* **109**, 10–24.

BURKE, R. E. 1967. Composite nature of the monosynaptic excitatory post-synaptic potential. *J. Neurophysiol.* **30**, 1114–1137.

BURKE, R. E. and NELSON, P. G. 1966. Synaptic activity in motoneurons during natural stimulation of muscle spindles. *Science* **151**, 1088–1091.

BURKE, W. 1957. Spontaneous potentials in slow muscle fibres of the frog. *J. Physiol.* **135**, 511–521.

BURNSTOCK, G. and HOLMAN, M. E. 1962. Spontaneous potentials at sympathetic nerve endings in smooth muscle. *J. Physiol.* **160**, 446–460.

COUTEAUX, R. 1955. Localization of cholinesterases at neuromuscular junctions. *Intern. Rev. Cytol.* **4**, 335–375.

COUTEAUX, R. 1963. The differentiation of synaptic areas. *Proc. R. Soc. B.* **158**, 457–480.

DALE, H. H., FELDBERG, W. and VOGT, M. 1936. Release of acetylcholine at voluntary motor nerve endings. *J. Physiol.* **86**, 353–380.

DEL CASTILLO, J. and KATZ, B. 1954a. The effect of magnesium on the activity of motor nerve endings. *J. Physiol.* **124**, 553–559.

DEL CASTILLO, J. and KATZ, B. 1954b. Quantal components of the end-plate potential. *J. Physiol.* **124**, 560–573.

DEL CASTILLO, J. and KATZ, B. 1954c. Changes in end-plate activity produced by pre-synaptic polarization. *J. Physiol.* **124**, 586–604.

DEL CASTILLO, J. and KATZ, B. 1954d. The membrane change produced by the neuromuscular transmitter. *J. Physiol.* **125**, 546–565.

DEL CASTILLO, J. and KATZ, B. 1955a. On the localization of acetylcholine receptors. *J. Physiol.* **128**, 157–181.

DEL CASTILLO, J. and KATZ, B. 1955b. Local activity at a depolarized nerve-muscle junction. *J. Physiol.* **128**, 396–411.

DEL CASTILLO, J. and KATZ, B. 1956a. Biophysical aspects of neuro-muscular transmission. *Progr. Biophys.* **6**, 121–170.

DEL CASTILLO, J. and KATZ, B. 1956b. Localization of active spots within the neuromuscular junction of the frog. *J. Physiol.* **132**, 630–649.

DEL CASTILLO, J. and KATZ, B. 1957a. La base 'quantale' de la transmission neuro-musculaire. In 'Microphysiologie comparée des éléments excitables.' *Coll. internat. C.N.R.S. Paris*, No. 67, 245–258.

DEL CASTILLO, J. and KATZ, B. 1957b. The identity of 'intrinsic' and 'extrinsic' acetylcholine receptors in the motor end-plate. *Proc. R. Soc. B.* **146**, 357–361.

DE ROBERTIS, E., PELLEGRINO DE IRALDI, A., RODRIGUEZ, G. and GOMEZ, C. J. 1961. On the isolation of nerve endings and synaptic vesicles. *J. biophys. biochem. cytol.* **9**, 229–235.

DODGE, F. A. JR. and RAHAMIMOFF, R. 1967. Co-operative action of calcium ions in transmitter release at the neuromuscular junction. *J. Physiol.* **193**, 419–432.

Douglas, W. W. 1966. Calcium-dependent links in stimulus-secretion coupling in the adrenal medulla and neurohypophysis. In: *Mechanisms of release of Biogenic Amines*, ed. von Euler, U. S., Rosell, S. and Uvnäs, B., pp. 267–290. Oxford: Pergamon Press.

Dudel, J. and Kuffler, S. W. 1961. The quantal nature of transmission and spontaneous miniature potentials at the crayfish neuromuscular junction. *J. Physiol.* **155**, 514–529.

Dudel, J. and Orkand, R. K. 1960. Spontaneous potential changes at crayfish neuromuscular junctions. *Nature, Lond.* **186**, 476–477.

Eccles, J. C. 1964. The Physiology of Synapses. Berlin: Springer, XI and 316 pp.

Eccles, J. C., Katz, B. and Kuffler, S. W. 1942. Effect of eserine on neuromuscular transmission. *J. Neurophysiol.* **5**, 211–230.

Eide, E., Fedina, L., Jansen, J., Lundberg, A. and Vyklický, L. 1967. Unitary excitatory postsynaptic potentials in Clarke's column neurones. *Nature, Lond.* **215**, 1176–1177.

Elmqvist, D. and Feldman, D. S. 1965. Spontaneous activity at a mammalian neuromuscular junction in tetrodotoxin. *Acta Physiol. Scand.* **64**, 475–477.

Elmqvist, D., Johns, T. R. and Thesleff, S. 1960. A study of some electrophysiological properties of human intercostal muscle. *J. Physiol.* **154**, 602–607.

Elmqvist, D. and Quastel, D. M. J. 1965. Presynaptic action of hemicholinium at the neuromuscular junction. *J. Physiol.* **177**, 463–482.

Fatt, P. and Katz, B. 1952. Spontaneous subthreshold activity at motor nerve endings. *J. Physiol.* **117**, 109–128.

Feng, T. P. and Li, T. H. 1941. Studies on the neuromuscular junction. XXIII. A new aspect of the phenomena of eserine potentiation and post-tetanic facilitation in mammalian muscles. *Chin. J. Physiol.* **16**, 37–56.

Furukawa, T., Sasaoka, T. and Hosoya, Y. 1959. Effects of tetrodotoxin on the neuromuscular junction. *Jap. J. Physiol.* **9**, 143–152.

Ginsborg, B. L. 1967. Ion movements in junctional transmission. *Pharmac. Rev.* **19**, 289–316.

Graham, J. and Gerard, R. W. 1946. Membrane potentials and excitation of impaled single muscle fibers. *J. cell. comp. Physiol.* **28**, 99–117.

Gray, E. G. and Whittaker, V. P. 1962. The isolation of nerve endings from brain: an electron microscopic study of cell fragments derived by homogenization and centrifugation. *J. Anat.* **96**, 79–88.

Hagiwara, S. and Tasaki, I. 1957. Demonstration of two stable potential states in the squid axon under tetraethylammonium chloride. *J. gen. Physiol.* **40**, 859–885.

Hagiwara, S. and Tasaki, I. 1958. A study on the mechanism of impulse transmission across the giant synapse of the squid. *J. Physiol.* **143**, 114–137.

Harvey, A. M. and MacIntosh, F. C. 1940. Calcium and synaptic transmission in a sympathetic ganglion. *J. Physiol.* **97**, 408–416.

HODGKIN, A. L. 1964. The Conduction of the Nervous Impulse. Liverpool University Press. 108 pp.

HODGKIN, A. L. and KEYNES, R. D. 1957. Movements of labelled calcium in squid giant axons. *J. Physiol.* **138**, 253–281.

HUBBARD, J. I. 1961. The effect of calcium and magnesium on the spontaneous release of transmitter from mammalian motor nerve endings. *J. Physiol.* **159**, 507–517.

HUBBARD, J. I. and SCHMIDT, R. F. 1963. An electrophysiological investigation of mammalian motor nerve terminals. *J. Physiol.* **166**, 145–165.

JENKINSON, D. H. 1957. The nature of the antagonism between calcium and magnesium ions at the neuromuscular junction. *J. Physiol.* **138**, 434–444.

KAO, C. Y. 1966. Tetrodotoxin, saxitoxin and their significance in the study of excitation phenomena. *Pharmac. Rev.* **18**, 997–1049.

KATZ, B. 1958. Microphysiology of the neuromuscular junction. *Johns Hopkins Hosp. Bull.* **102**, 275–312.

KATZ, B. 1962. The transmission of impulses from nerve to muscle, and the subcellular unit of synaptic action. *Proc. R. Soc. B.* **155**, 455–477.

KATZ, B. 1966. Nerve, Muscle and Synapse. McGraw-Hill, New York. IX and 193 pp.

KATZ, B. and MILEDI, R. 1963. A study of spontaneous miniature potentials in spinal motoneurones. *J. Physiol.* **163**, 389–422.

KATZ, B. and MILEDI, R. 1965a. Propagation of electric activity in motor nerve terminals. *Proc. R. Soc. B.* **161**, 453–482.

KATZ, B. and MILEDI, R. 1965b. The measurement of synaptic delay, and the time course of acetylcholine release at the neuromuscular junction. *Proc. R. Soc. B.* **161**, 483–495.

KATZ, B. and MILEDI, R. 1965c. The effect of calcium on acetylcholine release from motor nerve terminals. *Proc. R. Soc. B.* **161**, 496–503.

KATZ, B. and MILEDI, R. 1965d. Release of acetylcholine from a nerve terminal by electric pulses of variable strength and duration. *Nature, Lond.* **207**, 1097–1098.

KATZ, B. and MILEDI, R. 1965e. The effect of temperature on the synaptic delay at the neuromuscular junction. *J. Physiol.* **181**, 656–670.

KATZ, B. and MILEDI, R. 1966. Input/output relation of a single synapse. *Nature, Lond.* **212**, 1242–1245.

KATZ, B. and MILEDI, R. 1967a. Tetrodotoxin and neuromuscular transmission. *Proc. R. Soc. B.* **167**, 8–22.

KATZ, B. and MILEDI, R. 1967b. The release of acetylcholine from nerve endings by graded electric pulses. *Proc. R. Soc. B.* **167**, 23–38.

KATZ, B. and MILEDI, R. 1967c. The timing of calcium action during neuromuscular transmission. *J. Physiol.* **189**, 535–544.

KATZ, B. and MILEDI, R. 1967d. A study of synaptic transmission in the absence of nerve impulses. *J. Physiol.* **192**, 407–436.

KATZ, B. and THESLEFF, S. 1957a. On the factors which determine the amplitude of the 'miniature end-plate potential'. *J. Physiol.* **137**, 267–278.

KATZ, B. and THESLEFF, S. 1957b. The interaction between edrophonium (tensilon) and acetylcholine at the motor end-plate. *Brit. J. Pharmacol.* **12**, 260–264.

KOELLE, G. B. and FRIEDENWALD, J. S. 1949. A histochemical method for localizing cholinesterase activity. *Proc. Soc. exp. Biol. (N.Y.)* **70**, 617–622.

KUNO, M. 1964. Quantal components of excitatory synaptic potentials in spinal motoneurones. *J. Physiol.* **175**, 81–99.

KUSANO, K., LIVENGOOD, D. R. and WERMAN, R. 1967. Tetraethylammonium ions: effect of presynaptic injection on synaptic transmission. *Science* **155**, 1257–1259.

LEWIS, P. R. and SHUTE, C. C. D. 1964. Demonstration of cholinesterase activity with the electron microscope. *J. Physiol.* **175**, 5-7P.

LILEY, A. W. 1956a. An investigation of spontaneous activity at the neuromuscular junction of the rat. *J. Physiol.* **132**, 650–666.

LILEY, A. W. 1956b. The quantal components of the mammalian end-plate potential. *J. Physiol.* **133**, 571–587.

LILEY, A. W. 1956c. The effects of presynaptic polarization on the spontaneous activity of the mammalian neuromuscular junction. *J. Physiol.* **134**, 427–443.

LILEY, A. W. 1957. Spontaneous release of transmitter substance in multi-quantal units. *J. Physiol.* **136**, 595–605.

LING, G. and GERARD, R. W. 1949. The normal membrane potential of frog sartorius fibers. *J. cell. comp. Physiol.* **34**, 383–396.

LOCKE, F. S. 1894. Notiz über den Einfluss physiologischer Kochsalzlösung auf die elektrische Erregbarkeit von Muskel und Nerv. *Zbl. Physiol.* **8**, 166–167.

MARTIN, A. R. 1955. A further study of the statistical composition of the end-plate potential. *J. Physiol.* **130**, 114–122.

MARTIN, A. R. 1966. Quantal nature of synaptic transmission. *Physiol. Rev.* **46**, 51–66.

MARTIN, A. R. and PILAR, G. 1964. Quantal components of the synaptic potential in the ciliary ganglion of the chick. *J. Physiol.* **175**, 1–16.

MASLAND, R. L. and WIGTON, R. S. 1940. Nerve activity accompanying fasciculation produced by prostigmine. *J. Neurophysiol.* **3**, 269–275.

MILEDI, R. 1960a. The acetylcholine sensitivity of frog muscle fibres after complete and partial denervation. *J. Physiol.* **151**, 1–23.

MILEDI, R. 1960b. Junctional and extra-junctional acetylcholine receptors in skeletal muscle fibres. *J. Physiol.* **151**, 24–30.

MILEDI, R. 1961. From nerve to muscle. *Discovery*, **22**, 442–450.

MILEDI, R. 1964. Electron-microscopical localization of products from histochemical reactions used to detect cholinesterase in muscle. *Nature, Lond.* **204**, 293–295.

MILEDI, R. 1966. Miniature synaptic potentials in squid nerve cells. *Nature, Lond.* **212**, 1240–1242.

MILEDI, R. 1967. Spontaneous synaptic potentials and quantal release of transmitter in the stellate ganglion of the squid. *J. Physiol.* **192**, 379–406.

MILEDI, R. and SLATER, C. R. 1963. A study of rat nerve-muscle junctions after degeneration of the nerve. *J. Physiol.* **169**, 23*P*.

MILEDI, R. and SLATER, C. R. 1966. The action of calcium on neuronal synapses in the squid. *J. Physiol.* **184**, 473–498.

MILEDI, R. and THIES, R. E. 1967. Post-tetanic increase in frequency of miniature end-plate potentials in calcium-free solutions. *J. Physiol.* **192**, 54–55*P*.

MOSHER, H. S., FUHRMAN, F. A., BUCHWALD, H. D. and FISCHER, H. G. 1964. Tarichatoxin—tetrodotoxin: a potent neurotoxin. *Science* **144**, 1100–1110.

NACHMANSOHN, D. 1959. Chemical and Molecular Basis of Nerve Activity. New York: Academic Press. 235 pp.

NARAHASHI, T., MOORE, J. W. and SCOTT, W. R. 1964. Tetrodotoxin blockage of sodium conductance increase in lobster giant axons. *J. gen. Physiol.* **47**, 965–974.

NASTUK, W. L. 1959. Some ionic factors that influence the action of acetylcholine at the muscle end-plate membrane. *Ann. N.Y. Acad. Sci.* **81**, 317–327.

NISHI, S. and KOKETSU, K. 1960. Electrical properties and activities of single sympathetic neurons in frogs. *J. cell. comp. Physiol.* **55**, 15–30.

OTSUKA, M. and ENDO, M. 1960. The effect of guanidine on neuromuscular transmission. *J. Pharmacol. exp. Therap.* **128**, 273–282.

PALADE, G. E. 1959. Functional changes in the structure of cell components. In: *Subcellular Particles.* ed. T. Hayashi. Washington, D.C.: American Physiological Society. pp. 64–83.

RANDIĆ, M. and STRAUGHAN, D. W. 1964. Antidromic activity in the rat phrenic nerve-diaphragm preparation. *J. Physiol.* **173**, 130–148.

REUBEN, J. P. and GRUNDFEST, H. 1960. Inhibitory and excitatory miniature postsynaptic potentials in lobster muscle fibers. *Biol. Bull.* **119**, 335–336.

RIKER, W. F. JR. 1966. Actions of acetylcholine on mammalian motor nerve terminal. *J. Pharmacol.* **152**, 397–416.

SHERRINGTON, Sir Charles. 1947. The Integrative Action of the Nervous System. Cambridge University Press. XXIV and 433 pp.

SMITH, U. and SMITH, D. S. 1966. Observations on the secretory processes in the corpus cardiacum of the stick insect, *Carausius morosus. J. Cell Sci.* **1**, 59–66.

TAKEUCHI, A. 1959. Neuromuscular transmission of fish skeletal muscle investigated with intracellular microelectrode. *J. cell. comp. Physiol.* **54**, 211–220.

TAKEUCHI, A. and TAKEUCHI, N. 1962. Electrical changes in pre- and post-synaptic axons of the giant synapse of *Loligo. J. gen. Physiol.* **45**, 1181–1193.

TAKEUCHI, A. and TAKEUCHI, N. 1964. The effect on crayfish muscle of iontophoretically applied glutamate. *J. Physiol.* **170**, 296–317.

THESLEFF, S. 1960. Supersensitivity of skeletal muscle produced by botulinum toxin. *J. Physiol.* **151**, 598–607.

Tomita, T. 1967. Current spread in the smooth muscle of the guinea-pig vas deferens. *J. Physiol.* **189**, 163–176.

Usherwood, P. N. R. 1961. Spontaneous miniature potentials from insect muscle fibres. *J. Physiol.* **169**, 149–160.

von Euler, U. S., Rosell, S. and Uvnäs, B. 1966. Mechanisms of Release of Biogenic Amines. Oxford: Pergamon Press, 482 pp.

Whittaker, V. P., Michaelson, I. A. and Kirkland, R. J. A. 1964. The separation of synaptic vesicles from nerve-ending particles ('synaptosomes'). *Biochem. J.* **90**, 293–303.

Author Index

Andersson-Cedergren, E., 42
Armstrong, C. M., 39

Barstad, J. A. B., 45
Bernard, C., 31, 32
Binstock, L., 39
Birks, R., 4, 11, 44
Blackman, J. G., 47
Bloedel, J., 39
Boyd, I. A., 13, 17–19
Braun, M., 25
Brooks, V. B., 11
Brown, G. L., 1
Buchwald, H. D., 31
Bullock, T. H., 36
Burgen, A. S. V., 11
Burke, R. E., 47
Burke, W., 47
Burnstock, G., 47

Couteaux, R., 2, 3

Dale, H. H., 1, 3
del Castillo, J., 1, 2, 7, 10–13, 16, 17, 20, 21, 42
de Robertis, E., 11, 15, 48
Dickens, F., 11
Dodge, F. A. Jr., 13
Douglas, W. W., 47
Dudel, J., 47

Eccles, J. C., 45, 46
Eide, E., 47
Elmqvist, D., 31, 42, 47
Endo, M., 44

Fatt, P., 1, 5, 6, 11, 13, 14, 20
Fedina, L., 47

Feng, T. P., 45
Feldberg, W., 1, 41
Feldman, D. S., 31
Fischer, H. G., 31
Friedenwald, J. S., 2
Fuhrman, F. A., 31
Furukawa, T., 31

Gage, P. W., 39
Gerard, R. W., 5
Ginsborg, B. L., 2, 47
Gomez, C. J., 11, 15, 48
Graham, J., 5
Gray, E. G., 48
Grundfest, H., 47

Hagiwara, S., 36, 39
Harvey, A. M., 13
Hodgkin, A. L., 31, 39
Holman, M. E., 47
Hosoya, Y., 31
Hubbard, J. I., 14, 20
Huxley, H. E., 4

Jansen, J., 47
Jenkinson, D. H., 13
Johns, T. R., 47

Kao, C. Y., 31
Katz, B., 2, 4, 6, 7, 9–14, 16, 17, 20–29, 32–39, 45–47
Keynes, R. D., 39
Kirkland, R. J. A., 11, 48
Koelle, G. B., 2
Koketsu, K., 47
Kuffler, S. W., 46, 47
Kuno, M., 47
Kusano, K., 39

Lewis, P. R., 4
Li, T. H., 45
Liley, A. W., 11, 13, 14, 17, 20, 44
Ling, G., 5
Livengood, D. R., 39
Llinás, R., 39
Locke, F. S., 13
Lundberg, A., 47

MacIntosh, F. C., 13
Masland, R. L., 45
Martin, A. R., 12–14, 17–19, 44
Michaelson, I. A., 11, 48
Miledi, R., 1, 2, 4, 8, 9, 11, 13, 14, 20, 22–39, 45–47
Moore, J. W., 31
Mosher, H. S., 31

Nachmansohn, D., 32
Narahashi, T., 31
Nastuk, W. L., 2
Nelson, P. G., 47
Nishi, S., 47

Orkand, R. K., 47
Otsuka, M., 44

Palade, G. E., 15
Pellegrino de Iraldi, A., 11, 15, 48
Pilar, G., 44

Quastel, D. M. J., 39, 42

Rahamimoff, R., 13
Randić, M., 45
Ray., C., 47

Reuben, J. P., 47
Riker, W. F. Jr., 46
Ringer, S., 13
Rodriguez, G., 11, 15, 48
Rosell, S., 48

Sasaoka, T., 31
Schmidt, R. F., 20, 25
Scott, W. R., 31
Sherrington, C. S., 1
Shute, C. C. D., 4
Slater, C. R., 11, 36
Smith, D. S., 15
Smith, U., 15
Straughan, D. W., 45

Takeuchi, A., 36, 47
Takeuchi, N., 36, 47
Tasaki, I., 36, 39
Thesleff, S., 9–11, 47
Thies, R. E., 14
Tomita, T., 47

Usherwood, P. N. R., 47
Uvnäs, B., 48

Vogt, M., 1
von Euler, U. S., 48
Vyklický, L., 47

Werman, R., 39
Whittaker, V. P., 11, 48
Wigton, R. S., 45

Zatman, L. J., 11

Subject Index

Acetylcholine, 1–3, 6–9, 32, 43, 45, 46, 48
  effect on membrane permeability of, 2, 45
  quantal or molecular release of, 43
  synthesis of, 42, 43
Acetylcholine receptors, 32, 46
All-or-none events, 9, 16, 26
Anti-esterases, 7, 9, 45, 46
Arthropod muscle, 47

Basement membrane, 4
Botulinum toxin, 11

Calcium, 13, 14, 33, 34, 36, 39, 47
Calcium conductance, 34, 35
Calcium pipette, 22, 23, 25, 29, 33, 34
Carbachol, 9
Catecholamines, 47
Choline acetyltransferase, 48
Choline esters, 8, 9
Cholinesterase, 2, 3, 7, 9, 21
Curare, 7, 8, 31, 32, 46

Decamethonium, 9
Digoxin, 44

Edrophonium, 7–9, 45
End-plate potential, 2, 9, 12, 31, 36
  ionic mechanism of, 2
  quantal components of, 12, 14, 17, 18, 25, 28, 29, 32
Exocytosis, 15

Focal recording, 5, 19–22, 24, 28

Glutamate, 47
Guanidine, 44

Hemicholinium, 42

Ionophoretic application, 5, 7–10, 27, 34, 45

Latency shift, 35, 36

Magnesium, 13, 14, 31, 33, 34, 36, 47
Membrane conductance, 12, 13
Micro-electrode, 5, 20
Miniature end-plate potentials, 5–11, 18, 20–22, 42–44, 47
  frequency of, 11, 12
  non-linear summation of, 12, 13
Miniature synaptic potentials, 36, 47

Motor end-plate *see* neuromuscular junction
Multiquantal release, 44
Muscle impulse, 1, 2

Neostigmine, 7, 9, 32
Nerve ending, 2–4, 11, 15, 16, 20–22, 24, 33, 46, 48
  action potential of, 22, 24, 25, 27–29
  degeneration of, 11, 46
  depolarization of, 12, 15
Nerve impulse, 1
  ion conductance change during, 31
  retrograde discharge of, 45, 46
Neuromuscular junction, 2–5, 20, 31, 32, 35, 46

Poisson's theorem, 17, 19, 28–30,
    43, 44
Postsynaptic folds, 3–5
Potassium, 45
Potassium conductance, 39
Procaine, 31

Quantal release, 9, 14–17, 26, 27,
    31, 34, 41–43, 46, 47
  probability distribution of, 26
Quantum, 41, 42

Schwann cells, 11
Smooth muscle, 47
Sodium, 33
Spinal neurones, 47
Spontaneous activity, *see* miniature
  end-plate potentials

Spontaneous inhibitory potentials,
    47
Squid, giant synapse of, 35–38
Sympathetic ganglia, 47
Synapse, 1, 47
  input/output relation of, 37, 39
Synaptic cleft, 3, 4
Synaptic delay, 24–26
  effect of temperature on, 27
Synaptic vesicles, 4, 5, 10, 15, 16,
    34, 42, 44, 48

Tetraethylammonium, 38, 39
Tetrodotoxin, 31–34, 37–39
Transducer action, 2
Transmitter release, suppression of,
    38, 39
Tubocurarine, *see* curare